AF598615

ADVANCES IN BIOCHEMICAL ENGINEERING

Volume 16

Managing Editor: A. Fiechter

With 82 Figures

Springer-Verlag
Berlin Heidelberg New York 1980

ISBN 3-540-09807-0 Springer-Verlag Berlin Heidelberg New York
ISBN 0-387-09807-0 Springer-Verlag New York Heidelberg Berlin

Library of Congress Catalog Card Number 72-152360
Printed in GDR

2152/3020-543210

Preface

Biochemical Engineering is an interdisciplinary science. Basic aspects of Biology, Biochemistry, Engineering and related fields are major components of this vast field to be applied to a number of industrial and public areas.

The aim of this series is to uncover the characteristic features of Biochemical Engineering and to report on current developments observed in the application of basic knowledge to this discipline. In earlier stages of Biochemical Engineering, emphasis was laid on processes involving microbes or parts of the latter. Even now this type of cells is by far the dominant representative. Consequently, most articles of this series are dealing with topics on Microbiology demonstrating the importance of bacteria, yeasts, fungi or actinomycetes. Also, volumes 10 and 12 on "Immobilized Enzymes" show that these biocatalysts are essentially won from microbes. Much progress has been made in plant science, and cancer research has initiated extensive investigations on animal and human cells. Human cells are of particular interest in view of future medical applications but large-scale cultivation methods are still limited. The particular features of animal cells create considerable nutritional and mechanical problems of developing efficient submerged cultures for the production of biochemicals or viruses. Much interest has also been paid to plant cells. Progress made in submerge cultivation of cells or tissues is quite impressive and indicates high potentialities for practical applications. It also shows that improved techniques may provide useful tools for the study of basic problems arising in metabolism, morphogenesis and development of plants. Volumes 16 and 18 of this series describe some of the relevant advances made in this interesting field of Biology. The topics covered by this collection on review articles give an impression of present and future applications of Biochemical Engineering to the production of pharmaceuticals and other biochemicals as well as valuable procedures for practical work in plant production. A real "Plant Engineering" is emerging from such activities of research and the classical Biochemical Engineer may learn that his own methods making use of bioreactors are replaced by other classical methods, e.g. those of agriculture. Nevertheless, it seems important for both sides to see how different fields are coming closer and areas of common interest become evident. Biotransformations, production of steroids or continuous cultivation are certainly familiar to the Biochemical Engineer. He will also take note of the rapidly growing progress made in genetic manipulation of plants or will extrapolate the results of enzyme regulation studies or lipid chemistry to potential applications in mass culture of plant cells. However, other articles like low-temperature storage of plant cells or the regeneration of virus-free plants through *in vitro* culture will more appeal to the agriculturist. Other endeavours like Embryogenesis, regulation of Morphogenesis or cell cycle studies are mandatory for future progress to be made in the basic knowledge of developmental Biology. Plant cell research is still a field of rather small scientific community. The recent death of a prominent member, Prof. H. E. Street from Leicester University, is a big loss and creates a gap which can hardly be filled in a short time. He was one of the leading experts in plant tissue culture and the first president of the "International Plant Tissue Culture Association". As a basically trained pharmacist he was active in Plant Physiology and significantly contributed to Plant Nutrition at Nottingham. He was professor of Botany at Swansea and later at Leicester, mainly as a plant physiologist who became interested in organ and cell culture. The symposium on "The Impact of Plant Tissue Culture on Industry and Agriculture" at the Calgary Congress of the IAPTC was dedicated to the unique contributions to plant tissue and cell cultures written by this prominent British scientist.

We would like to express the feelings of high appreciation of the entire community of plant cell culture specialists in our series and dedicate Volume 16 in memory of our fine colleague and great scientist Prof. H. E. Street.

Zürich, January 1980 A. Fiechter

Contents

Continuous Culture of Plant Cells Using the Chemostat Principle

G. Wilson
Department of Botany University College Dublin
Belfield, Dublin 4, Ireland

Whereas plant cells have been grown as batch cultures for many years the application of continuous culture methods is relatively very new. This article considers how the design of a plant cell chemostat is influenced by the special growth and morphological characteristics of plant cells. The achievement of steady-state growth enabled the application of kinetic models earlier developed with microorganisms. The advantages of chemostat culture in enabling the distinction between effects of growth rate and effects of growth-limiting substrate allows the identification of factors influencing cell composition and cell metabolism with a precision unobtainable in batch culture. Some further potential applications of steady-state studies with plant cells are outlined.

1 Introduction

The application of microbiological methods and experience to form an emerging new biotechnology for cultured plant cells has featured in recent years[1]. The technique of continuous culture using the chemostat principle was developed originally with bacteria[29] for the purpose of enabling the control of growth in defined equilibrium conditions. The potential of the method, both as a research tool and as a production tool outlined by Tempest[49], led to its application to fungi, animal, and more recently to plant cell culture. The continuous culture of plant cells has been facilitated by the advent of relatively finely dispersed and homogeneous cell suspensions capable of growth in chemically defined media. The aim of the present review is to indicate the practicality of the method and to provide an idea of the scope of its application to plant cell growth and metabolism.

2 Characteristics of Plant Cells

2.1 Structure

Despite the observation[34)] that cultured cells will 'be a new group of microorganisms' it is pertinent to outline briefly the special morphological and growth characteristics of plant cells because these impose special requirements on the design of bioreactors and continuous culture systems. Cells from a great many plant species have now been grown successfully in liquid suspension as batch cultures[46,41)]. Whenever growing suspensions of plant cells have been examined they have been found to contain both cell aggregates and free cells. The relative proportion of free cells and the size range of aggregates varies considerably between different cultures. In the most homogeneous cultures at the present time, e.g. *Morinda citrifolia*, approximately[62)] 60% of the cells are either free cells or clusters of two cells. In many cell suspensions however cell aggregates up to 2 mm diameter are not uncommon. The origin of these aggregates is the result of the pattern of cell division[11)]. Characteristically the cultured plant cell divides to form an internal cross wall. Repeated divisions of this kind result in the formation of a cell cluster in which the cells are held together by their shared internal walls. In batch culture, at the end of the phase of cell division, continued cell enlargement may enable the separation of free cells from the aggregates. As a result, the degree of cell aggregation within a batch culture may vary considerably according to the growth phase. Plant cells do not aggregate in the way some animal cells do.

Plant cells in suspension culture have a thin unspecialised cell wall (0.2 μ to 0.6 μ across) which encloses the protoplast. Small perforations in the wall with plasmodesmata furnish protoplasmic continuity between adjacent cells. Besides containing an elaborate membrane system characteristic of a eukaryotic cell the protoplast also contains numerous large vacuoles often transversed by cytoplasmic strands showing active streaming movements. The protoplast is osmotically active and in the absence of a wall, or if the wall becomes broken in the hypotonic conditions of normal culture media, will take up water, expand and burst. Therefore one of the functions of the wall in maintaining cellular integrity is to provide an opposite pressure (wall pressure) on the protoplast.

Individually, plant cells are relatively much larger than bacteria or fungi but even within a single culture there may be a wide variety of shapes and sizes ranging from nearly spherical to approximately cylindrical. The linear dimensions of cultured plant cells are usually within the range 20–40 μ diameter and 100–200 μ long. Thus on a volume basis plant cells are some 100,000–200,000 times larger than bacteria. By comparison therefore mass transfer processes in plant cells are large and metabolic rates several times lower. The relatively long doubling time of cultured plant cells is usually between 20–40 h, exceeding that of bacteria[32)] by a factor of 60–100.

2.2 Culture Conditions

The basis of all nutrient media for plant cells is a mixture of inorganic salts together with a carbon (and energy) source which is usually sucrose or glucose. Apart from a few cell strains most plant cells require this basic medium to be supplemented with certain plant growth regulators, the phytohormones (e.g. IAA, NAA, 2.4:D and kinetin); vitamins (e.g. thiamin, nicotinic acid); amino acids (casein hydrolysate) and the sugar alcohol inositol. For cell division plant cells require aerobic conditions and so, additionally, need aeration either by shaking or bubbling. Details of these fundamental requirements are previously reviewed[46]. Most plant cell culture media have been generally designed to enable maximum increase in biomass, usually assessed by fresh weight or dry weight[30].

3 Growth and Metabolic Patterns of Plant Cells in Batch Culture

The common practice of initiating a closed batch culture with an inoculum of non-dividing cells produces a characteristic sigmoid growth curve in respect of cell density. This is adequately summarised by the model growth curve shown by King *et al.*[22] (Fig. 1).

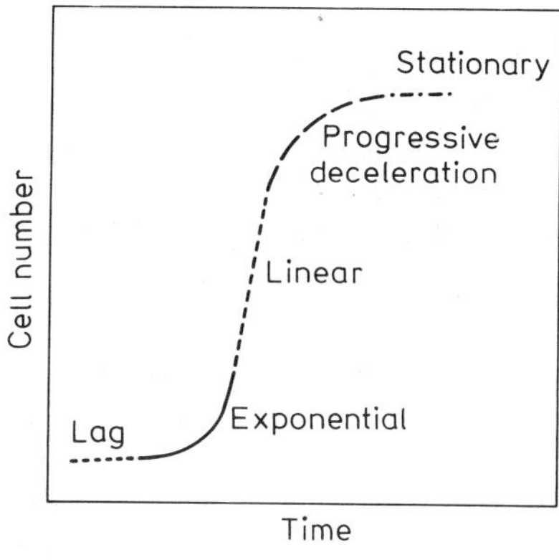

Fig. 1 Model growth curve illustrating growth phases during batch culture of a plant cell suspension (after King *et al.*[22])

Apart from differences in the duration of different phases such a curve appears to describe experimentally obtained measurements for several plant cell cultures, e.g. *Acer*[10], *Galium mollugo*[55], *Rosa*[31]. Two points are of interest about this typical growth curve for plant cells. Firstly, as has been noted by Street[45], the total overall increase in cell number (or total dry weight) is usually only about 10–15 fold, or about 3–4 generations, commonly representing an increase in cell number from 0.2×10^6 ml^{-1} to 3×10^6 ml^{-1}. Secondly, the phase of strictly exponential growth is quite a small proportion of the whole 'growth cycle' occupying only 2–3 generations of cell division.

The term "growth cycle" used to describe the overall succession of events following the inoculation of cells into a fixed volume of nutrient medium was inherited from microbiology. The growth cycle was rationalised for bacteria by Monod, who resolved the exponential phase as being a constant growth phase flanked on either side by two transient states, the initial 'lag' phase in which cells adapted to new nutrient conditions and adjusted to growth at a maximal rate, and the 'stationary' phase which followed after nutrient exhaustion. Comparison with the 'growth cycle' of higher plant cells reveals, not unexpectedly, many similarities. Nevertheless there is now an impressive body of evidence which shows that in batch culture plant cells undergo a continually changing pattern of biosynthetic activity. In several different plant cell cultures it has been observed from growth analysis that although a constant exponential rate of cell division is achieved for a short time other parameters of cell metabolism do not increase concomitantly and there is no period of steady-state growth in which the relative cell concentrations of all metabolites and enzymes are constant. This absence of 'balanced' growth[4)], resulting in marked changes in cell composition, has been shown in *Acer* cells for several aspects of cell metabolism including protein and nucleic acid synthesis, respiration rate, the activity of several enzymes e.g. phenylalanine-ammonia-lyase, glucose-6-phosphate dehydrogenase, nitrate reductase, as well as ethylene production[22)]. Similar results have also been obtained with *Rosa* cells[31)], the enzymes of phenylpropanoid biosynthesis in soybean cells[12, 13)] and anthraquinone biosynthesis in *Galium mollugo* cells[55)] (Fig. 9). Because of the different rates of synthesis of different cell metabolites in relation to the cell division rate characteristic and significant changes in cell composition occur throughout the batch growth cycle. These sequential changes develop during the initial induction from a non-dividing to a dividing condition, before an overall steady metabolism is achieved cell division becomes retarded. Very significant reduction in nutrient concentrations follow soon after inoculation. With *Acer* cells almost all the phosphate and 90% of the medium nitrogen is taken up before even one generation of cell growth is achieved[54)]. Clearly in this case cell division cannot be said to accompany a progressive balance of nutrient uptake. King *et al.*[22)] suggested that, in view of this, the slowing down and eventual cessation of cell division may be the result of the utilisation of endogenous levels of metabolites accumulated in the cells much earlier in the growth cycle rather than more directly by a related depletion of a primary nutrient from the medium. Following an exceptionally detailed analysis of growth and metabolism of batch cultured *Ipomoea* cells Rose *et al.*[43, 44)] concluded that the growth cycle was sufficiently more complicated than that of bacterial cultures that one should not ordinarily make comparisons. With the wisdom acquired from recent research with continuous cultures it can readily be seen that the lack of steady-state metabolism in batch culture does not represent a fundamental difference between plant cells and microorganisms but it is more likely that the combination of the relatively small amount of exponential growth supported in batch culture together with the long generation time leads to a high inertia of the system[3)] and retards the development of steady-state conditions.

So it must be recognised that plant cells in a batch culture are in a state of continual physiological change resulting from the environmental change occurring in a closed system. This complexity presents an obstacle to the precise understanding not only of how the concentration of nutrient components determine patterns of growth and metabolism, but also how metabolic rates and rates of cell division are fundamentally related, if at all. In the past, many observed kinetic relationships may well turn out to be artifacts of the batch culture system.

4 Continuous and Semi-continuous Culture Systems

4.1 Terminology

In the literature reference is made to several forms of continuous or semi-continuous culture systems and it is important to distinguish between them as each may have special characteristics. Essentially plant cell cultures may be either 'open' or 'closed' systems. The system is closed if some part of the culture cannot enter or leave it: an example is the batch culture which contains a fixed volume of nutrient medium. Normally in batch cultures of plant cells growth ceases when an essential nutrient becomes depleted. Thus the growth rate of the cells must tend towards zero.

Several forms of 'open' culture systems have been described, the feature common to all is that, at intervals, additional fresh medium is added and samples of culture removed. By this means the growth of the culture may be maintained continuously. Culture systems of this kind in which addition of fresh medium and withdrawal of samples are made at arbitrarily defined intervals are generally known as semi-continuous cultures. The special forms of 'open' continuous culture are the chemostat and turbidostat systems in which fresh medium is continuously added to a fixed volume of culture. The essential feature of these two methods, which is the subject of this review, is that cell division takes place in steady-state conditions at a defined and uniform rate.

There is, therefore, a clear functional and biological distinction between batch culture, semi-continuous culture and steady-state continuous cultures. Nevertheless experience and advances in the bioengineering aspects of large scale bioreactors, many of which function as batch cultures or semi-continuous cultures, inevitably relate also to continuous culture methods.

4.2 Merits of the Continuous (Chemostat) Culture System

In batch culture the biosynthetic activity of cultured plant cells varies with the growth rate and substrate availability. The study of factors influencing growth and metabolism of plant cells is facilitated by three special features of chemostat culture[38, 49].

a) The chemostat allows control of the growth rate with no change to the environment other than the concentration of the growth limiting substrate.
b) The chemostat can be used to fix a steady-state growth rate while the environment is altered. For example similar growth rates may be maintained under either phosphate or nitrate limitation. This is useful in order to distinguish between effects of change of growth rate and change of the limiting nutrient on cell metabolism.
c) The chemostat can be used to maintain substrate-limited growth while at the same time maintaining a constant environment. This is in contrast to the characteristics of a closed batch culture in which substrate-limited growth is attained only transiently and is accompanied by a changing growth rate and environment.

The chemostat technique can therefore offer two distinct advantages: it can extend the range of conditions possible in a culture and it can be used to overcome some of the difficulties of interpreting the complex patterns of growth that always occur in a batch culture. These two features may enable a better understanding to be made of the regulation of metabolism in cultured plant cells.

5 Application of the Continuous (Chemostat and Turbidostat) Culture System to Plant Cells

The development of continuous culture methods using the chemostat and turbidostat principle for plant cells was stimulated by the realisation of the obstacles to further progress presented by batch culture methods in identifying the factors influencing the rates of cell division and the control of specific aspects of cellular metabolism. Undoubtedly progress in this field has been strongly guided and influenced by previous knowledge gained in microbiological systems. Monod[27, 28)] had shown how microbial growth could be formulated in terms of the growth yield (Y), the specific growth rate (μ_{max}) and the concentration of the growth-limiting nutrient (s). From this analytical approach followed the chemostat type of continuous flow culture, initiated by Monod[28)], Novick and Szilard[33)] and later developed by Herbert *et al.*[15)]. By enabling steady-state growth the chemostat culture method offered a precise means of elucidating relations between an organism and its environment.

The mathematical model, originally developed for microorganisms, providing the theory for the development of steady-state growth in continuous culture systems has been published several times[15, 24, 9)] and will not be derived in full here. This theory has been shown to account reasonably well for the observed growth of microorganisms, both bacteria[15)] and mycelial fungi[35)], and has also been used to serve as a model describing plant cell growth in continuous culture[46)].

5.1 Outline Theory

The conditions provided by a chemostat enable cell division to take place in steady-state conditions at a defined and uniform rate under the control of a chosen limiting nutrient. Under these conditions growth becomes balanced[4)] in the sense that the cell composition remains constant, characteristic of the division rate and the limiting nutrient.

In its simplest form the chemostat consists of a culture vessel into which fresh medium is added continuously to a fixed volume of growing culture. The composition of the input medium is adjusted so that a single nutrient is limiting. The cells grow at a rate determined by the medium input rate and the culture overflows. Once steady-state growth is achieved the rate of cell division is exactly equal to the rate of dilution by fresh medium. Ultimately the growth rate is determined by the concentration of the growth limiting nutrient in the chemostat vessel.

The key to the operation of the chemostat lies in the way the cell division rate (μ) is related to the concentration of the growth-limiting substrate in the medium. In a medium in which all the essential components are in excess except one (the growth-limiting substrate), the rate of cell division was shown by Monod[29)] to be related to the limiting substrate concentration in the manner shown in Fig. 2.

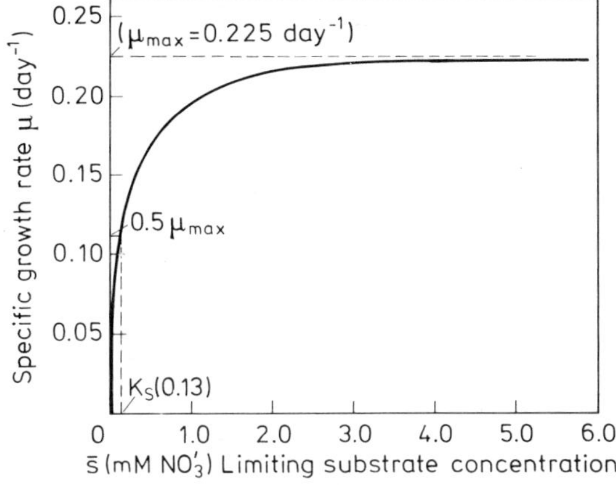

Fig. 2 Relationship between μ (specific growth rate) and $\bar{s}$ (steady-state limiting substrate concentration) in a chemost at according to Eq. 1 (see text). Parameters: $\mu_{max} = 0.225\ day^{-1}$, $K_s = 0.13$ mM (NO_3') (data from King *et al.*[22] obtained for *Acer* cells)

For several microorganisms the experimental curves obtained are fitted reasonably well by the equation suggested by Monod —

$$\mu = \mu_{max} \frac{s}{K_s + s} \qquad \text{(Eq. 1, Fig. 2)}$$

where: s = limiting substrate concentration in the culture
K_s = substrate concentration enabling half the maximum growth rate
μ = growth rate at limiting substrate concentrations
μ_{max} = maximum growth rate.

In a batch culture the nutrient concentrations are reduced as the cells grow, consequently the growth rate (μ) decreases and both eventually fall to zero. In a chemostat however the substrate concentration is maintained by the constant addition of fresh medium and in consequence the rate of cell division is fixed at some point on the μ/s curve, Fig. 2. In operation, provided the dilution rate is set to a suitable value the rate of cell division adjusts itself to a steady-state in which the cell density and the limiting substrate concentration remain constant indefinitely. Thus by adjustment of the dilution rate (D) different steady-state growth rates can be maintained. The dilution rate is a measure of the tendency for cells to be washed out from the culture and is expressed as:

$$D = \frac{\text{input flow rate}}{\text{culture volume}} \frac{(\text{ml h}^{-1})}{(\text{ml})} .$$

Thus (D) the dilution rate has the same units (h^{-1}) as the growth rate (μ) and in steady state conditions:

$$D = \mu .$$

In a steady-state a dynamic equilibrium is established for the limiting nutrient which may be equated as follows:

Equilibrium substrate concentration = Input of substrate — Output of substrate — Consumption of substrate

Expressed algebraically:

$$\frac{ds}{dt} = DS_R - D_s \frac{\mu x}{Y}$$

where: S_R = limiting substrate concentration in the input medium
s = equilibrium limiting substrate concentration in the chemostat culture
x = steady-state cell density
Y = yield coefficient $\frac{\text{(cells formed)}}{\text{(substrate utilised)}}$

The relationship between the limiting substrate concentration S_R and the steady-state cell density (x) obtained was originally suggested by Monod[29] as follows:

$$Y = \frac{x}{S_R - s}.$$

In summary, Monod's model of microorganism growth enables its definition in relation to the nutrient environment in terms of three factors; μ_{max}, K_s and Y.

5.2 Factors Influencing Plant Cell Chemostat Design

It should be emphasised that steady-state continuous culture in chemostats and turbidostats under defined limiting nutrient conditions has so far been restricted to *Acer* (Sycamore) and *Galium* (Greater Hedgebedstraw) cells. Nevertheless a wider experience of the growth of many different plant cell suspensions in relatively large scale bioreactors contributes much to our general knowledge of bioengineering principles for plant cells[17, 60, 51].

The physical characteristics of plant cells present specific problems for chemostat or turbidostat culture. Fundamental to the design of any chemostat is the requirement for complete culture mixing and, in addition, an overflow system which besides maintaining a constant culture volume allows no discrimination between cells and medium. Whereas the aggregates of many plant cell cultures are sufficiently large to settle out of suspension quickly, the relative fragility and sensitivity to shearing forces of plant cells[51, 52] generally prohibits the combination of high speed mechanical stirring and baffles commonly used for microorganisms. Furthermore, the simple weir types of overflow (as used on the Porton Pot[49] does not function indiscriminately with *Acer* cells (Wilson, G. unpublished) due to cell aggregates being diverted by the edge of the lip of the overflow pipe. Resulting from this general experience several novel methods of culture mixing and overflow design have appeared in chemostats especially designed for plant cells. The problem of overflow is overcome in one design[57] by using an electronic level detector coupled to an interval timer which energises a solenoid activated valve that allows a volume of culture to be released. In the Kurz[25] design the overflow is controlled by an electronically timed pump. A simpler method in which the culture overflow is continuously exhausted along with the aeration air is used successfully with finely dispersed cultures[54]. Most designs have been constructed from laboratory glassware (Figs. 3 & 4), but the use of a modified L.K.B. 'Ultraferm' has been described[8]. Successful turbidostat operation has been made possible by the addition of a photoelectric density sensing device fitted to an

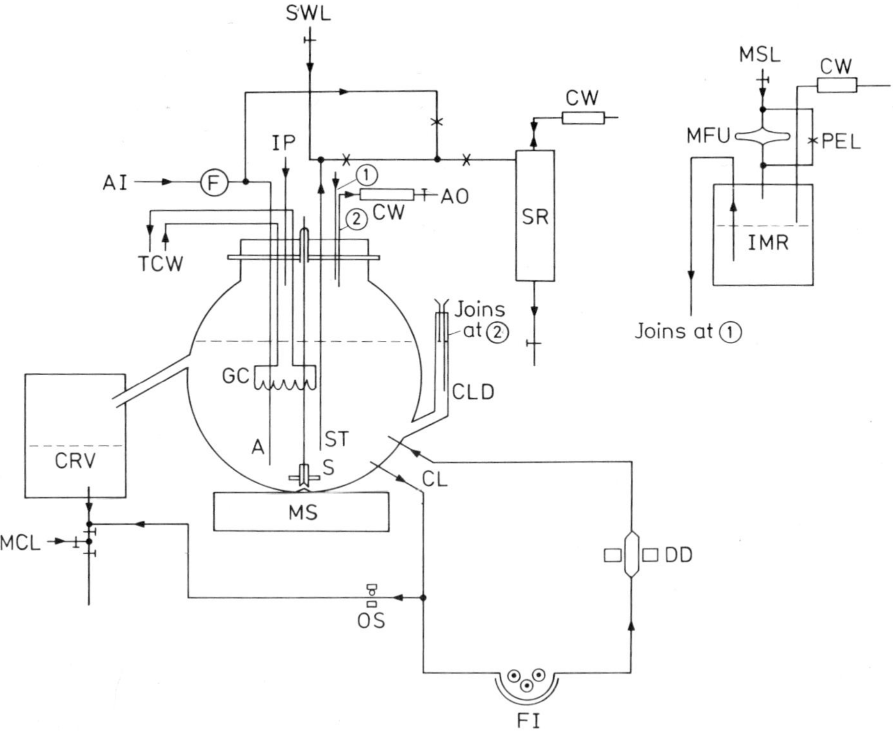

Fig. 3 Flow diagram for 4L chemostat continuous culture system (Wilson, S. B., *et al.*[57])
Key: A = aerator (No. 2 sintered); AI = air inlet; AO = air outlet; CL = circulation loop; CLD = constant level device; CRV = culture receiving vessel; CW = cotton wool filter; DD = culture density detector; F = air filter (Microflow Ltd.); FI = flow inducer; GC = glass coil (for temperature control); IMR = intermediate medium reservoir; MCL = mercuric chloride wash line; MFU = medium filter unit; MS = magnetic stirrer motor; MSL = new medium supply line; MS = magnetic stirrer motor; OS = outlet solenoid valve (through which culture harvested in response to constant level device); PEL = pressure equalising loop; SR = sample reservoir; SWL = sterile water line; ST = sample tube; TCW = temperature controlled water

external circulating loop[57]. Turbidometric control using an optical fibre is described by Bligny[2].

Generally a gentle form of culture mixing using air-bubble-generated turbulence is favoured, sometimes in conjunction with a low speed suspended magnetic paddle: 500 ml min^{-1}, 260 rpm[57]. Use of a magnetic paddle alone to obtain efficient mixing is reported to cause cell breakage[56] and these workers used a finger-action stirrer. In the Kurz column design a unique, pulsed air-stream is used to produce large individual air bubbles that are injected at the base of a cylindrical culture vessel. This is reported to improve cell separation. Use of an inverted conical design[54] enables good mixing by air bubbles alone at the relatively high flow rate of 2 l min^{-1} (2 vvm). Under these conditions, when used as a batch culture, the growth pattern obtained is identical to that obtained in shake flasks. The oxygen demand of plant cell cultures is not high[17,19,52], although this clearly depends on the culture steady-state density. There is therefore no need for high-shear stirring to increase gas transfer. The use of

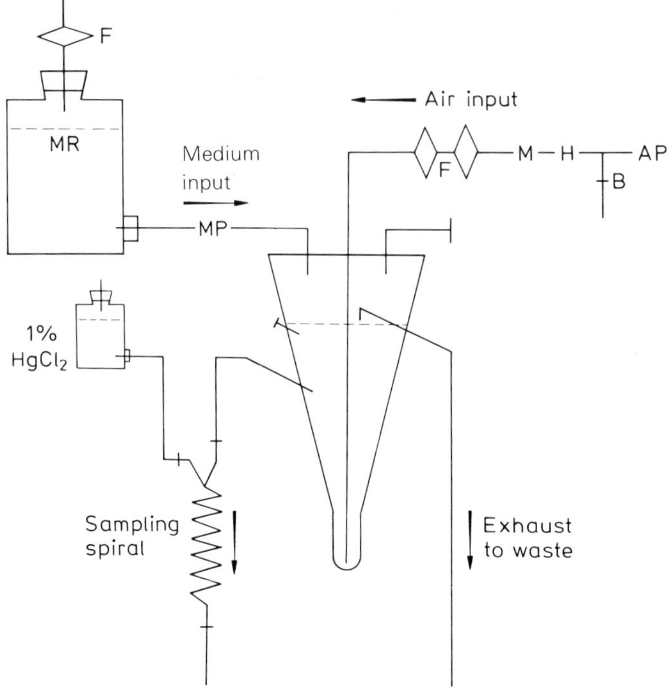

Fig. 4 Flow diagram of 1L chemostat system (Wilson[54])
Key: MR = medium reservoir; MP = medium pump; F = air filter; M = air flow rate meter; H = humidifier; AP = air pump; B = air bleed

a sintered sparger is reported[57] but a plain ended tube is used satisfactorily[54]. There is a tendency for plant cells to grow tenaciously inside the submerged tip of the aeration tube, and also on spargers, which can lead to troublesome blockages.

The steady-state culture density can be adjusted by use of a suitable concentration of growth-limiting nutrient in a chemostat and by appropriate optical density setting in a turbidostat. High cell densities place increasing demands on the stirring and aeration systems and give rise to the possibility of blockages in the overflow and excessive cell accretion to internal surfaces. Both of these problems cause distortion to steady-state conditions. Generally plant cell chemostats are best operated at low cell densities e.g. 4 g dry wt. l^{-1}. This is substantially less than the final density that can be obtained in batch culture and thus problems of mixing highly viscous cultures can be avoided[20, 52]. At low operational cell densities foaming is relatively light rendering the addition of antifoam unnecessary. Although plant cells do cause changes in the culture medium pH, changes in continuous culture appear relatively small[54] Fig. 5)[19], and no attempt has been made to automatically maintain a constant pH.

The relatively long mean generation times of plant cells[32] places an emphasis on troublefree and reliable equipment to facilitate the time period necessary for the establishment of steady-state conditions. Culture periods of 30—60 days are not unusual. To avoid disturbance of the steady-state the culture volume needs to be chosen with regard to the size and frequency of sampling. The culture volume used for *Acer* cells (doubling time 80 h)[57] is about 4 l, but a smaller culture volume (700 ml) is used for fast growing *Galium* cells (doubling time 35 h)[55].

5.3 The Establishment of Steady-state Conditions in Continuous Culture

The term 'steady-state' describes the situation in which equilibrium conditions exist in a culture. This is defined in terms of a constant cell density, cell composition and nutrient environment. In a chemostat these conditions are progressively obtained after inoculation following the setting of the dilution rate to a suitable fixed value. With plant cells several authors have noted that different parameters of cell metabolism require varying lengths of time for equilibration[23, 54]. Obviously the time required for equilibration will depend upon the nature and size of the inoculum and the initial dilution rate. Nevertheless, some seven generations were required for equilibration of dry weight and limiting nutrient concentration (t_d = 43 h) with *Acer* cells[54] (Fig. 5), eight generations with *Galium* (t_d = 25 h) for dry weight and anthraquinone concentration (Wilson, G., unpublished) and one and a half to two generations for dry weight (t_d = 220 h) *Acer*[57]. Equilibration of amino-N pool sizes in *Acer* cells were reported to require several months after stabilisation of the cell density in *Acer* cells (t_d = 125 h)[21]. The response time for a more defined transition, that of cell density changes in moving from one steady-state to another induced by a reduction in

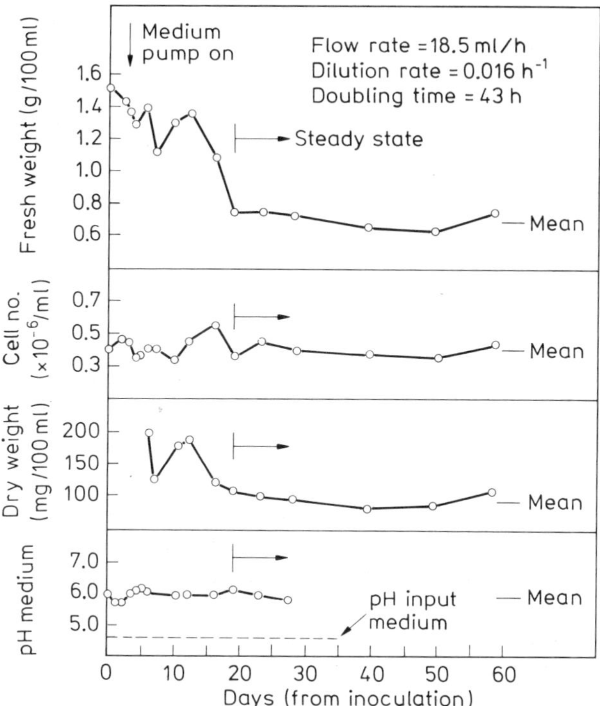

Fig. 5 Development of a steady state from inoculation for several growth parameters (*Acer* cells) (from Wilson[54])

limiting nutrient concentration (constant dilution rate), was about ten generations ($t_d = 117$ h)[24]. Here the time course of the change closely followed a theoretically predicted curve based upon Monod's microorganism model (Fig. 6). This example suggests that the number of generations required for equilibrium is not significantly different from that of microorganisms. However, the extensive time required for equilibration illustrates the high inertia of plant cells to their environment and confirms the unlikelyhood of steady-state development in a batch culture in which growth equivalent to 3–4 generations occurs in a continually changing environment.

5.4 Tests of Theory: the Growth Parameters Y, μ_{max} and K_S

5.4.1 Yield Coefficient

The yield coefficient (Y) relates the amount of growth obtained per unit of substrate and is therefore important as a quantitative measure of the nutrient requirements of a cell culture. It is often assumed that, in steady-state substrate-ilimited growth, the amount of limiting substrate per cell is the minimum compatible with growth. Although the yield coefficient (Y) was initially conceived as a growth constant with microorganisms there are now several widely recognised examples of variation in response to changing growth rate[39].

With plant cells there have been no systematic studies on the yield coefficient although values have been calculated for some fixed growth rates (Table 1).

Table 1. Yield coefficient (Y) values obtained for cultured plant cells

Cell culture	Limiting nutrient	Yield coefficient	Doubling time	Reference
		10^6 cells/µmole	(h)	
Galium	Phosphate (PO_4^-)	3.95	40	Wilson[55]
Galium	Phosphate (PO_4^-)	2.47	25	Wilson[55]
Acer	Phosphate (PO_4^-)	3.47	182	Wilson[54]
Acer	Phosphate (PO_4^-)	1.28	36	Wilson[54]
Acer	Nitrate (NO_3^-)	0.263	109	King[24]
Acer	Glucose	0.039	109	King[24]

Expressed on a dry weight basis the yield coefficient (*Acer*) for glucose can be calculated from data of King[24] as approximately 0.37 g dry wt/g glucose. Kato, A. *et al.*[19] obtained values between 0.5 g g^{-1}–0.36 g g^{-1} for tobacco cells. These values compare closely with that obtained for *Penicillium*, 0.433 g g^{-1} [35], and also for other microorganisms.

The response, expressed as change in equilibrium cell density, to a change in the concentration of the limiting nutrient (NO_3^-) in the input medium was found to follow a time course predicted by Monod kinetics[23] (Fig. 6). In this study, at a constant dilution rate, the predicted curve was calculated using a yield coefficient value obtained previously during steady-state growth. These results, which apply equally both in the case of an increase or decrease in nitrate concentration,

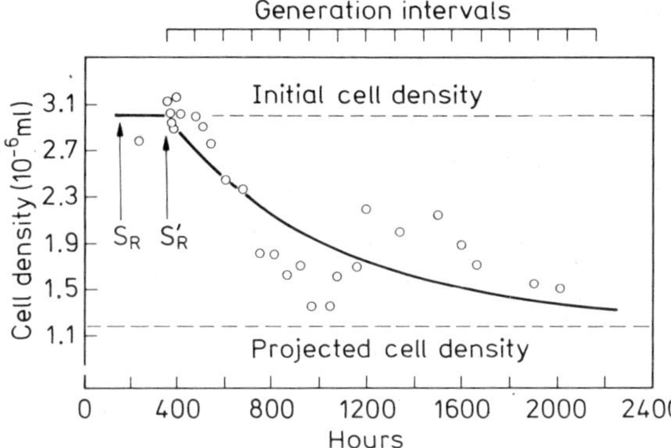

Fig. 6 Test of theory: Decrease in cell density in a chemostat following reduction of the input concentration of the limiting nutrient (NO_3'). The superimposed curve shows the predicted response based upon Monod kinetics.

$$x = S_R\left[Y\frac{S'_R}{S_R} - Y\left(\frac{S'_R}{S_R} - 1\right)e^{-Dt}\right]$$

$S_R = 8.75 \times 10^{-3}$ M NO_3'
$S'_R = 3.5 \times 10^{-3}$ M NO_3'
Dilution rate (D) = 0.142 d^{-1}
Mean Generation Time = 117 h
(after King[24])

provide an informative link between the growth kinetics of plant cells in suspension culture and those of microorganisms.

5.4.2 Specific Growth Rate

Stable, steady-state growth rates have been maintained over the range of doubling times 70 h–332 h for *Acer* cells[22]. These correspond to close to μ_{max} and to about 20% μ_{max}. The critical dilution rate (D_c) in chemostat studies generally agrees well with μ_{max} observed with the same cell line in batch culture[54] and also in turbidostat culture[57]. With *Galium* cells stable steady-states between doubling times 25 h to 55 h are reported[55]. In this case the critical dilution rate in chemostat culture (approximately equivalent to a doubling time of 25 h) is greater than the μ_{max} in batch culture (equivalent to a doubling time of 35 h). This is probably the result of wall growth, a phenomenon also recognised in microorganism culture. There are, so far, no reports of a minimum growth rate[37].

5.4.3 Substrate Saturation Constant: K_S

The achievement of steady-state growth of plant cells in chemostatic conditions has enabled the determination of some nutrient saturation constants K_S (Table 2). This requires measurement of steady-state limiting substrate concentrations at different defined growth rates and then substituting these values in Monod's equation (Fig. 2). In practice limiting nutrient concentrations are usually very low[61] and additionally, may change substantially before a sample can be filtered of cells and assayed. The K_S for nitrate has been calculated using a Lineweaver-Burk plot method[22]. Here the linearity of the double reciprocal plot forms a convenient test of validity

Table 2. Substrate saturation constant (K_S) values obtained for plant cells

Cell culture	Substrate	K_S (mM)
Acer (Sycamore)	NO_3^-	0.13
Acer (Sycamore)	Glucose	0.5
Acer (Sycamore)	PO_4^-	0.032

of Monod's equation relating substrate concentration to specific growth rate. It is noteworthy that the application of this method did not give a linear plot for glucose as limiting nutrient for *Acer* cells[22]. The estimation of K_S for phosphate[53] was calculated only from a single growth rate and is therefore certainly only approximate. It has been observed that these values are larger than K_S concentrations for similar substrates for some microorganisms, but in view of the limited data available it is probably premature to ascribe any general significance to this comparison. The identification of kinetics of the control of rates of cell division by limiting nutrients in cultured plant cells opens up speculation on the possibility of similar regulation in intact plants where most cells are not dividing and in which rapidly dividing cells are normally spatially defined in meristems.

5.5 Effect of Growth Rate on Cell Composition and Metabolism

All the data that refers to changes in cell composition and physiology with respect to growth rate has been obtained under nitrogen (probably nitrate) limited conditions, and with *Acer* cells[22, 7, 61]. The range of growth rates studied extends approximately from 0.06–0.215 d^{-1} (equivalent doubling times 277 h–77 h) and Fowler 7 reports a steady-state at the slow doubling time of 410 h. Plant cells grown at different steady growth rates differ significantly in their composition and metabolism. As yet, however, there has been no evidence to suggest the formation of stable morphogenetic states analogous to spore formation which occurs at low steady-state growth rates in some fungi, e.g. *Geotrichum*[42]. King, P. J.[22] found that increasing growth rate was accompanied by a decline in cell volume and cell dry

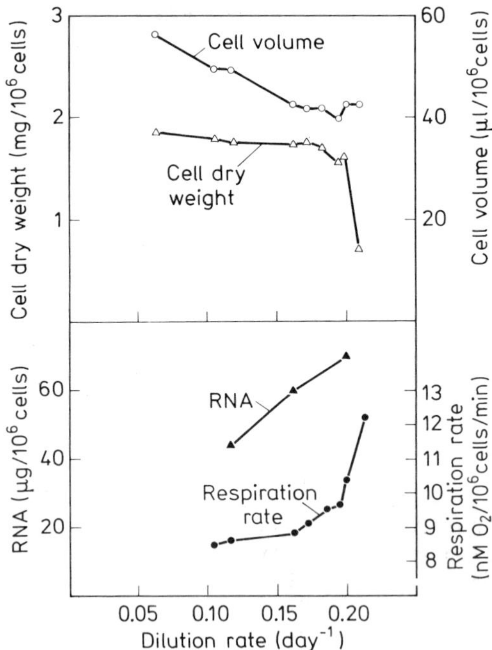

Fig. 7 Effect of growth rate (dilution rate) on cell composition and physiological activity in steady-state conditions (after King *et al.*[22])

weight but by marked increases in cell RNA and cell respiration rate (Fig. 7). Although similar, transient changes have also been observed to occur in batch culture, loosely associated with changing growth rate the significance of this steady-state data is that the cell composition, or physiological state is directly related to a defined growth rate and nutrient environment.

Studies on enzyme activity in the assimilative pathways of nitrogen and carbohydrate metabolism in chemostat culture have shown that stable steady-state levels of activity are achieved[7, 61)]. The mean steady-state activity of some N-assimilative enzymes e.g. nitrate reductase showed only small changes at different growth rates as did also cellular levels of amino acid pools. These results are consistent with nitrate being the growth limiting nutrient and are in marked contrast to the relatively much larger, transient, changes in nitrate reductase activity which occur in batch culture (Fig. 10). Undoubtedly the achievement of steady-state enzyme activities in defined environmental conditions will facilitate studies on the regulation of metabolic pathways in plant cells.

5.6 Effects of Limiting Nutrient on Cell Composition

Once steady-state conditions become established, although the division rate is exponential, cells grow under a specific from of substrate limitation and all other nutrients are in excess. It is quite clear, even from the few results obtained so far that the nature of the limiting nutrient significantly affects cell composition (Tables 3, 4). It is important, in order to obviate effects of growth rate, to compare treatments in which cells are grown at the same dilution rate.

Table 3. Steady-state data obtained at similar growth rates showing effect of limiting substrate on cell composition (*Acer* cells)

Limiting nutrient	Phosphate[a]	Nitrate[b]	Urea[c]
Dilution rate (d^{-1})	0.138	0.14	0.13
Cell dry wt. ($mg \times 10^{-6}$ cells)	4.9	1.6	5.1
Cell fresh wt. ($mg \times 10^{-6}$ cells)	42	16	48.2
Cell protein ($\mu g \times 10^{-6}$ cells)	—	341	1154

[a] from Wilson[53)]; [b] from Young[61)]; [c] from King[21)]

Table 4. Steady-state cell composition of *Acer* cells in A. nitrate limited conditions and B. following a change to glucose limiting conditions (see also Fig. 8) (after King[24)])

	A Nitrate	B Glucose
Dilution rate (d^{-1})	0.15	0.15
Cell dry weight ($mg \times 10^{-6}$ cells)	1.84	0.55
Cell volume ($\mu l \times 10^{-6}$ cells)	53	33
Cell starch ($\mu g \times 10^{-6}$ cells)	228	87
Cell protein ($\mu g \times 10^{-6}$ cells)	427	369

The reaction of plant cells to the nature of the limiting nutrient is illustrated by the cell compositional changes following alteration from nitrate to glucose limited growth[23)]. In this case, following a period of steady-state nitrate limited growth, a change to glucose limited growth was established by reducing the glucose concentration of the input medium (from 20 g l^{-1} to 5 g l^{-1} (Fig. 8)). The dilution rate was kept constant. After a transient period of about five generations a new steady-state developed in which the cell composition was markedly different, particularly in respect of reduced cell dry weight and cell starch levels (Table 4). Consequently protein occupied an increased proportion of the cell dry weight.

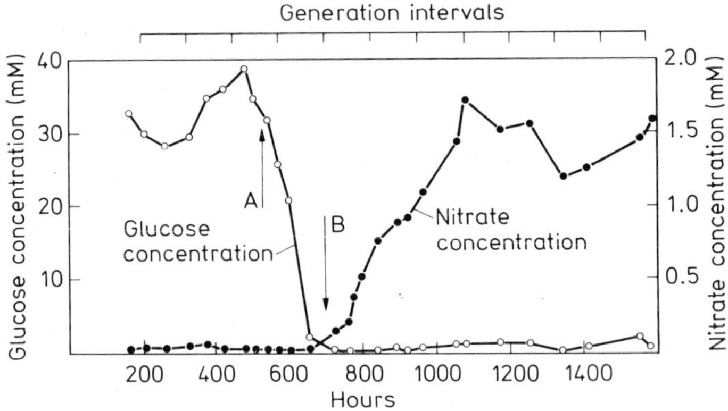

Fig. 8 Changes in nutrient concentrations following an alteration from nitrate to glucose limiting conditions
A: Concentration of glucose in the input medium reduced from 0.11 M to 0.0275 M after 536 h of nitrate limited growth
B: Glucose concentration in chemostat reaches steady-state growth limiting concentration. Excess nitrate begins to appear in the medium.
Dilution rate = 0.132 d^{-1}
Mean generation time = 126 h
(after King, P. J.[24)])

The relatively big reduction (30%) in cell dry weight following the change to glucose limitation is only partly due to the reduction in starch content. Interest is attached to the observation that the difference may be due to reduction in the wall fraction of the cells. The increased starch content and cell dry weight of *Acer* cells under nitrate limited conditions may be analagous to results obtained with *Torula utilis*[14)] and *E. coli*[16)] both of which accumulate the polysaccharide glycogen under conditions of nitrogen limitation (ammonia) in which the carbon source was in excess, whereas in glucose limiting conditions this did not occur. In *Acer* cells, however, it is significant that the cell compositional differences were only observed following a transition from nitrate to glucose limitation and were not found when cells were grown from inoculation in glucose limited conditions. It is unclear as yet how differences in earlier culture conditions can affect the later development of cell composition in steady-state conditions but this may be of some importance if further examples become defined.

Whether the phytohormones can be regarded in the same way as inorganic nutrients from the point of view of the kinetics of substrate-limited growth in chemostat continuous culture is at present unknown. This question is of some interest since growth and metabolism of cultured plant cells are strongly influenced by relatively low concentrations of phytohormones[5)]. The effects of making hormones growth-rate limiting have not been investigated but there is evidence that these might be dramatic. Using continuous culture the effect of total 2.4:D withdrawal from a steady-state *Acer* chemostat culture[23)] was to progressively reduce the rate of cell division. This resulted in the reduction in cell density due to 'wash out'. These results indicate an absolute 2.4:D requirement for cell division. Although no steady-state was obtained in this experiment there were increases in cell volume, cell protein and cell dry weight. Significantly a reduced cell viability accompanied 2.4:D withdrawal, in marked contrast to either nitrate or phosphate limitation[10)]. This may raise complications if 2.4:D limited chemostat culture is attempted. Using a turbidostat to maintain a constant cell density (fixed by optical density feedback to the medium input pump) enabled the establishment of prolonged steady-states at several different low 2.4:D concentrations. In these treatments the mean cell division rate was maintained close to μ_{max} so that it appears that 2.4:D was not growth-rate limiting. Nevertheless, in contrast to the cell composition obtained at the highest input 2.4:D concentration $4 \cdot 5 \times 10^{-6}$ M, at the lowest 2.4:D concentration (0.5×10^{-6} M) a marked enhance-

Table 5. Effect of high (A) and low (B) concentration of 2:4-Dichlorophenoxy acetic acid on the culture and cell composition of Acer cells in steady-state growth (after King[23)])

	2:4-D Treatment	
	A	B
Input 2:4-D conc. (10^{-6} M)	4.5	0.5
Steady-state spent medium 2:4-D conc. (10^{-6} M)	2.32	0.15
Cell aggregation (% culture dry weight)		
Less than 300 μm	30.6	20.7
300–1000 μm	64.8	65.6
More than 1000 μm	4.8	13.7
Lignin		
μg per 10^6 cells	22.1	105.2
% dry wt.	0.63	3.18
Total soluble phenolics		
Gallic acid equivalents μg per 10^6 cells	7.47	12.00
Coumaric acid		
μg per 10^6 cells	0.118	0.383

ment of phenolic metabolism was observed (Table 5). Steady-state measurements of 2.4:D in the spent medium showed that these responses occurred between 3.5 and 1.0×10^{-7} M. Accompanying the differences in chemical composition were increases in cellular aggregation and changes in cell morphology[58)]. A degree of structural and chemical cellular heterogeneity was found within the aggregates. Some centrally located cells accumulated large amounts of electron-dense material (possibly polyphenolic) and possessed walls with bands of thickening alternating with thinner wall containing plasmodesmata. These morphological changes were tentatively interpreted as the result of an imperfectly executed pathway of xylem differentiation. Whether the increased cell aggregation provides the necessary microenvironment for these compositional changes or, alternatively, whether aggregation results from the observed changes in wall structure is, as pointed out by the author, uncertain. Clearly however the response to low 2.4:D conditions is not uniform within the cell population as a whole.

5.7 Comparison between Batch and Chemostat Cultured Cells

The effect of a steady-state environment and growth rate obtained in chemostat culture on cell composition is clearly seen in contrast to that obtained in batch culture (Figs. 9, 10).

In the case of *Galium mollugo*[55)] steady-state values of cell protein, cell dry weight and cell anthraquinone were significantly lower than the transient peak values obtained in batch culture. Similar results were obtained for nitrate reductase activity (per mg protein) in *Acer* cells[61)]. Taking the three cell parameters for *Galium* together, cells in steady-state growth have a cell composition unlike cells at any point on the batch growth cycle. In particular, in chemostat culture, the cell anthraquinone levels are between 7 and 30 times lower than in batch culture. This suppression of anthraquinone biosynthesis may reflect the particular growth rate that was studied, or the specific effect of phosphate as the growth limiting nutrient. The chemostat culture technique offers the advantage that these two possibilities may be independently investigated.

These results raise the question, previously posed[22, 7, 46)] as to whether by choice of a suitable dilution rate it would be possible to maintain permanently in steady-state continuous culture cells having a composition or physiological state equivalent to a particular point in the batch growth cycle. Such a possibility might be of value for obtaining maximum productivity of a particular desired enzyme or secondary metabolite which in batch culture accumulated only transiently, such as in the case of anthraquinone synthesis at day 10–11 (Fig. 9). Deeper consideration of the contrasting events in batch culture and steady-state chemostat culture is sufficient to convince one that there is unlikely to be a close analogy between these two culture systems. If the reasons for the transient changes in cell composition in batch culture are the direct result of continuing, but out of step, changes in growth rate and product synthesis arising out of a changing environment it seems improbable that such transient physiological states could be maintained in the unique steady-state environmental conditions of the chemostat. In batch culture, at any point in time, the cellular physiological state is changing influenced not only by a changing environment but also by earlier physiological states which occurred in a different environment. This assessment, reached earlier in relation to microorganisms[47, 36, 40)] seems even more

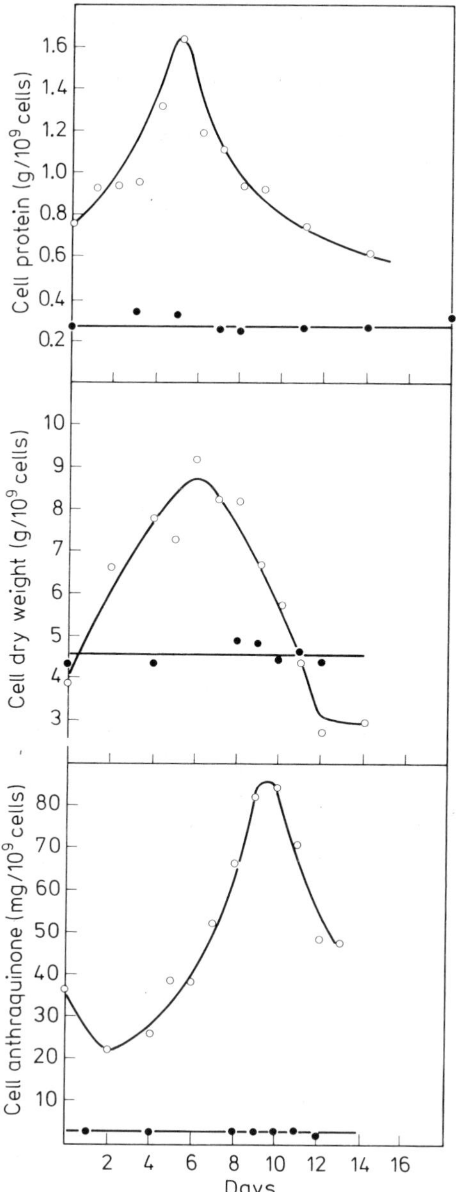

Fig. 9 Changes in cell composition during batch culture (○—○) and steady-state chemostat culture (●—●), *in Galium mollugo* cells (from Wilson and Marron[55])

likely to be the case for plant cells because of their relatively high 'inertia' to the environment. The results obtained under phosphate limited conditions (Fig. 9) clearly support this view and emphasise that by using a chemostat to obtain different steady-state environments one can expect to produce a range of cultures containing phenotypically different cells some with properties which are never expressed in batch culture.

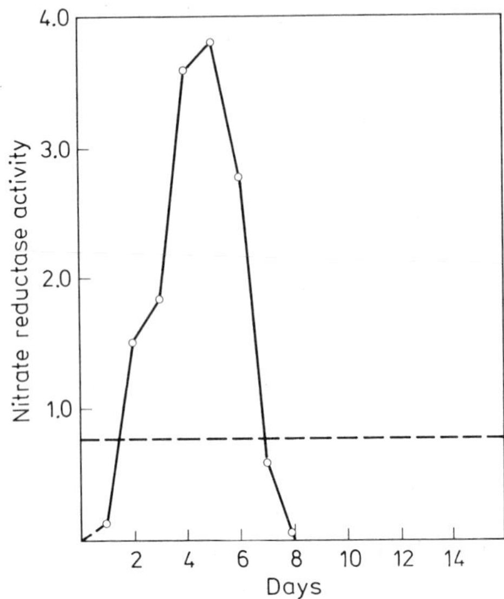

Fig. 10 Change in nitrate reductase activity (μM/min/mg Protein) during batch culture (○—○) and in continuous chemostat culture (— — — —) of *Acer* cells (after Young[61])

5.8 Biomass Production Using Large-scale Continuous Culture Methods

Both continuous and semi-continuous culture methods have been advocated for biomass production on account of the advantages of maintaining cells in the growth phase and thereby avoiding production delays due to the "lag" phase and early low density growth. Additionally these procedures minimise the costs of repeated sterilisation and the time required for reinoculation of successive batch cultures.

The theoretical relative increase in output productivity of a chemostat culture over that of repeated batch cultures is: —

$$\frac{\text{Output (chemostat)}}{\text{Output (batch)}} = \ln\left(\frac{x_m}{x_o}\right) + \frac{0.693\, t_a}{t_d} \qquad \text{(Pirt}^{38)}\text{)}.$$

Where x_m = maximum biomass concentration (batch)
x_0 = inoculum biomass concentration (batch)
t_a = delay time between successive exponential phase cycles in batch culture
t_d = fastest doubling time.
Making a rough approximation:

Taking $\frac{x_m}{x_0} = 10$, $t_a = 7$ days, $t_d = 1.5$ days.

Output chemostat = 5.5 × output batch culture. Clearly the relatively long mean doubling time for plant cells (t_d) reduces the gains in output productivity of continuous culture. The value of t_a for batch culture might be reduced by adopting a more frequent sub-culture regime.

Pioneering large-scale continuous cultures aimed at assessing biomass productivity of both single[19] and two stage[32] bioreactors show that relatively high productivity is possible. The single stage process (1500 l) produced up to 3.82 g dry wt. $l^{-1}\ d^{-1}$

($t_d = 40$ h) although problems due to inefficient stirring were reported. The two stage process (each bioreactor volume = 35 l) produced 6.9 g dry wt. l^{-1} d^{-1}, at a mean doubling time of 31 h.

6 Outlook

It is clear from the foregoing that the chemostat method markedly simplifies the culture system and facilitates the understanding between the plant cell and its environment. Up to now the growth of plant cells has been discussed in terms derived from bacteria and fungi with which it still largely conforms. Kinetic data for plant cells is still too meagre to make the assumption that microorganism kinetics will be found totally applicable. It may well prove that the points at which plant cell behaviour diverge from the familiar model will eventually be shown to be the most significant in the environmental regulation of cell division and metabolism.

The development of a wide range of finely dispersed cell suspensions obtained from different plants capable of growth in chemically defined media enables a more general application of chemostat and turbidostat steady state methods to growth in defined conditions. This is long overdue as there is a danger of too much extrapolation from *Acer* cell data.

Several applications to subjects of current interest can be envisaged. Clearly further tests of theory with a variety of limiting substrates over a wide range of growth rates would prove fundamentally rewarding. It may be that the spectrum of cell composition observed in batch culture is only a small fraction of that obtainable in substrate limited conditions. Systematic determination of Yield (Y) values showing their relationship with growth rate and with different substrates would provide an understanding into the quantitative utilisation of nutrients, enabling in the case of a growth limiting energy source the derivation of a value for the maintenance energy requirements of plant cells[38)].

Attempts to establish natural growth regulating compounds, the phytohormones, as growth limiting substrates obviously presents a unique challenge to plant cell biologists without exact parallel in microbiology. Although most cultured plant cells have an absolute requirement for various hormones[59)] the relationship between cell division rate and hormone concentration is unclear[22, 23, 26)]. Preliminary investigations will be needed to provide further kinetic data on this important point. However steady-state studies, by defining the stoichiometry of hormonal uptake and cell division, could provide a system advantageous to the study of binding sites.

The developing interest in the secondary metabolites produced by cultured plant cells raises the question of the nature of the controlling factors influencing patterns of biosynthesis in batch culture[1, 63, 55)]. In batch culture kinetic studies have shown that product biosynthesis is in some species coincident with growth and in others the product is only formed as the growth rate declines[48)]. This raises the question as to how product biosynthesis in plant cells is related to growth rate, if at all, or whether product synthesis is more influenced by changing nutrient concentrations. Several authors advocate the application of steady-state methods to the study of secondary product biosynthesis[3, 46)]). Clearly chemostat studies can contribute here by enabling a distinction between growth rate and substrate limitation[56)]. Only in this way can unequivocal conclusions be reached on the cause and effect relationships affecting plant secondary metabolite formation.

Lack of information relating to plant cells forces one to turn to experience with microbial secondary metabolite regulation[6, 38]. The nature of the limiting nutrient can effect secondary product biosynthesis quite markedly although the underlying mechanisms are not completely understood. For example penicillin production is inhibited by excess glucose and the biosynthesis of some antibiotics is stimulated by phosphate limitation. Limitation by biotin enables glutamine overproduction, as the result of increased membrane permeability, in *Corynebacterium* bacterium species. Citric acid production is stimulated by trace element limitation in *Aspergillus niger*. However, in practice, the selection of suitable controlling factors cannot be predicted and improvements in productivity are obtained by screening of different environmental conditions.

It would clearly be of value if the application of chemostat continuous culture methods enabled an understanding of the factors influencing plant cell secondary metabolism. Such fundamental information may be a valuable contribution to the developing subject of a biotechnology for higher plant cells.

Acknowledgements
I wish to acknowledge financial support from the Bundesministerium für Forschung und Technologie, Bonn, West Germany and the National Science Council, Ireland.

7 References

1. Barz, W., Reinhart, E., Zenk, M. H. (Eds.): Plant Tissue Culture and its Biotechnological Application. Proc. 1st Intern. Congr. on Medic. Plant, Res., Sect. B. Proc. Life Sciences, Berlin: Springer 1977
2. Bligny, R.: Growth of Suspension Cultured *Acer pseudoplatanus* L. cells in Automatic culture units of large volume. Plant Physiol. **59**, 502 (1977)
3. Butenko, R. G.: The cultivation of higher plant cells in suspension culture. Akad. Nauk. SSSR Ser. Biol. *5*, 697 (1977)
4. Campbell, A.: Synchronisation of Cell Division. Bact. Rev. *21*, 263 (1957)
5. Constabel, F., Shyluk, J. P., Gamborg, O. L.: The effect of hormones on anthocyanin accumulation in cell cultures of *Haplopappus gracilis*. Planta *96*, 306 (1971)
6. Demain, A. L.: Cellular and environmental factors affecting the synthesis and excretion of metabolites. J. Appl. Chem. Biotechnol. *22*, 345 (1972)
7. Fowler, M. W., Clifton, A.: Activities of enzymes of carbohydrate metabolism in cells of *Acer pseudoplatanus* L., maintained in continuous (chemostat) culture. Eur. J. Biochem. *45*, 445 (1974)
8. Fowler, M. W.: Continuous (chemostat) culture of plant cells using the L.K.B. Ultraferm fermentation system. L.K.B. Application Note 252 (1976)
9. Fowler, M. W.: Growth on cell cultures under chemostat conditions. In: Plant Tissue Culture and its Biotechnological Application. Barz, W., Reinert, E., Zenk, M. H. (Eds.), 253. Berlin: Springer 1977
10. Gould, A. R., Bayliss, M. W., Street, H. E.: Studies on the growth in culture of plant cells. XVII. Analysis of the cell cycle of asynchronously dividing *Acer pseudoplatanus* L. cells in suspension cultures. J. exp. Bot. *25*, 468 (1974)
11. Henshaw, G. G., Jha, K. K., Mechta, A. P., Shakeshaft, D. J., Street, H. E.: Studies on the growth in culture of plant cells. I. Growth patterns in batch propagated suspension cultures. J. exp. Bot. *17*, 362 (1966)

12. Hahlbrock, K.: Correlation between nitrate uptake, growth and changes in metabolic activities of cultured plant cells. In: Tissue Culture and Plant Science. Street, H. E. (Ed.), 363. London & New York: Academic Press 1974
13. Hahlbrock, K.: Regulatory aspects of phenylpropanoid biosynthesis in cell cultures. In: Plant Tissue Culture and its Bio-technological Application. Barz, W., Reinhart, E., Zenk, M. H. (Eds.), 95. Berlin: Springer 1977
14. Herbert, D.: The chemical composition of microorganisms as a function of their environment. Symp. Soc. Gen. Microbiol. *11*, 391 (1961)
15. Herbert, D., Elsworth, R., Telling, R. C.: The continuous culture of bacteria: a theoretical and experimental study. J. Gen. Microbiol. *14* (8), 501 (1956)
16. Holme, T.: Continuous culture studies on glycogen synthesis in *Escherichia coli* B. Acta Chem. Scand. *11*, 763 (1957)
17. Kato, K., Shiozawa, Y., Yamada, A., Nishida, K., Noguchi, M.: A jar fermenter culture of *Nicotiana tabacum* cell suspension. Agr. Biol. Chem. *36*, 899 (1972)
18. Kato, A., Shimizu, Y., Nagai, S.: Effect of initial K_La on the growth of tobacco cells in batch culture. J. Ferment. Technol. *53* (10), 744 (1975)
19. Kato, A., Kawazoe, S., Iijima, M., Shimizu, Y.: Continuous culture of tobacco cells. J. Ferment. Technol. *54* (2), 82 (1976)
20. Kato, K., Kawazoe, S., Yoshihiko, S.: Viscosity of the broth of tobacco cells in suspension culture. J. Ferment. Technol. *56* (3), 224 (1978)
21. King, J.: Growth characteristics of *Acer pseudoplatanus* L. cells grown in chemostat conditions in the presence of urea alone as a source of nitrogen. Plant Science Letters *6*, 409 (1976)
22. King, P. J., Mansfield, K. J., Street, H. E.: Control of growth and metabolism of cultured plant cells. Can. J. Bot. *51*, 1807 (1973)
23. King, P. J.: Studies on the growth in culture of plant cells. XX. Utilisation of 2,4-Dichlorophenoxyacetic acid by steady-state cell cultures of *Acer pseudoplatanus* L. J. exp. Bot. *27*, 100, 1053 (1976)
24. King, P. J.: Studies on the growth in culture of plant cells. XXII. Growth limitation by nitrate and glucose in chemostat cultures of *Acer pseudoplatanus* L. J. exp. Bot. *28* (102), 142 (1977)
25. Kurz, W. G. W.: A chemostat for growing higher plant cells in single cell suspension cultures. Exp. Cell Res. *64*, 476 (1971)
26. Leguay, J. J., Guern, J.: Quantitative effects of 2,4-Dichlorophenoxyacetic acid on growth of suspension cultured *Acer pseudoplatanus* cells. Plant Physiol. *56*, 356 (1975)
27. Monod, J.: Reserches sur la croissance des cultures bactériennes. Paris: Hermann & Co., 1942
28. Monod, J.: The growth of bacterial cultures. Ann. Rev. Microbiol. *3*, 371 (1949)
29. Monod, J.: La technique de culture continuee. Théorie et application. Paris: Ann. Inst. Pasteur *79*, 390 (1950)
30. Murashigi, T., Skoog, F.: A revised medium for rapid growth and bioassays with tobacco tissue cultures. Physiologia Pl. *15*, 743 (1962)
31. Nash, D. T., Davies, M. E.: Some aspects of growth and metabolism of Pauls Scarlet rose cell suspensions. J. Exp. Bot. *23*, 75 (1972)
32. Noguchi, M., Matsumoto, T., Hirata, Y., Yamamoto, K., Katsuyama, A., Kato, A., Azechi, S., Kato, K.: Improvements of growth rates of plant cell cultures. In: Plant Tissue Culture and its Biotechnological Application. Barz, W., Reinhart, E., Zenk, M. H. (Eds.), 85. Berlin: Springer 1977
33. Novick, A., Szilard, L.: Description of the chemostat. Science N.Y. *112*, 715 (1950)
34. Nickell, L. G., Tulecke, W.: Submerged growth of cells of higher plants. J. Biochem. Microbiol. Tech. and Eng. *2*, 287 (1960)
35. Pirt, S. J., Callow, D. S.: Studies of the growth of *Penicillium chrysogenum* in continuous flow culture with reference to Penicillin production. J. Appl. Bacteriol. *23* (1), 87 (1960)
36. Pirt, S. J.: Steady state processes for cell growth and product formation. Microbial Physiology and Continuous Culture 3rd Int. Symp., p. 162 H.M.S.O. 1967
37. Pirt, S. J.: Prospects and problems in continuous flow culture of microorganisms. J. Appl. Chem. Biotechnol. *22*, 55 (1972)

38. Pirt, S. J.: Principles of microbe and cell cultivation. Oxford: Blackwell Scientific Publications 1975
39. Postgate, J. R., Hunter, J. R.: The survival of starved bacteria. J. Gen. Microbiol. *29*, 233 (1962)
40. Powell, E. O.: The analogy between batch and continuous culture. Microbiol Physiology and Continuous Culture 3rd Int. Symp., p. 209 H.M.S.O. 1967
41. Puhan, Z., Martin, S. M.: The industrial potential of plant cell culture. Progress in Industrial Microbiology. Hockenhall, D.J.D., Clowes, W. (Eds.), Vol. *9*, p. 13. London, Colchester, Beccles 1971
42. Robinson, P. M., Smith, J. M.: Morphogenesis and Growth kinetics of *Geotrichum candidum* in continuous culture. Trans. Br. mycol. Soc. *66* (3) 413–420 (1976)
43. Rose, D., Martin, S. M., Clay, P. P. F.: Metabolic rates for major nutrients in suspension cultures of plant cells. Can. J. Bot. *50*, 1301 (1971)
44. Rose, D., Martin, S. M.: Parameters for growth measurement in suspension cultures of plant cells. Can. J. Bot. *52*, 903 (1973)
45. Street, H. E.: Applications of cell suspension cultures. In: Applied and Fundamental Aspects of Plant Cell, Tissue and Organ Culture. Reinert, J., Bajaj, Y.P.S. (Eds.), p. 649. Berlin: Springer 1976
46. Street, H. E.: (Ed.) Plant Tissue and Cell Culture. Oxford: Blackwell Scientific Publications 1977
47. Sikyta, B., Slezak, J., Herold, M.: Continuous chlortetracycline fermentation in continuous cultivation of microorganisms. Malek, I., Beran, K., Hospodta, J. (Eds.). Prague, Nakladatelstvi Ceskoslovonske Akademie Ved. 1964
48. Tabata, M.: Recent advances in the production of medicinal substances by plant cell cultures. In: Plant Tissue Culture and its Bio-technological Application. Barz, W., Reinhart, E., Zenk, M. H. (Eds.), p. 3. Berlin: Springer 1977
49. Tempest, D. W.: The place of continuous culture in microbiological research. Adv. Microbiol. Physiol. *4*, 223 (1970)
50. Verma, D., Van Huystee, R.: Derivation characteristics and large scale culture of a cell line from *Arachis hypogeae* L. cotyledons. Expl. Cell Res. *69*, 402 (1971)
51. Vogelmann, H., Bishof, D., Pape, D., Wagner, F.: Some aspects on mass cultivation. In: Production of Natural Compounds by Cell Culture Methods. Alfermann, A. W., Reinert, E. (Eds.). Muenchen: Gesellshaft fur Strahlen und Umweltforschung mbH, 1978
52. Wagner, F., Vogelmann, H.: Cultivation of plant tissue cultures in bioreactors and formation of secondary metabolites. In: Plant Tissue Culture and Its Biotechnological Application. Barz, W., Reinhart, E., Zenk, M. H. (Eds.), p. 245. Berlin: Springer 1977
53. Wilson, G.: The nutrition and differentiation of cells of *Acer pseudoplatanus* L. in suspension culture. Ph. D. thesis, University of Birmingham, England 1971
54. Wilson, G.: A simple and inexpensive design of chemostat enabling steady-state growth of *Acer pseudoplatanus* L. cells under phosphate limited conditions. Ann. Bot. *40*, 919 (1976)
55. Wilson, G., Marron, P.: Growth and anthraquinone biosynthesis by *Galium mollugo* L. cells in batch and chemostat culture. J. exp. Bot. *29* (111), 837 (1978)
56. Wilson, G.: Growth and product formation in large-scale and continuous culture systems. Frontiers in Plant Tissue Culture, b. Thorpe, T., (Ed.), p. 169. Pub. Int. Assn. Plant Tissue Culture 1978
57. Wilson, S. B., King, P. J., Street, H. E.: Studies on the growth in culture of plant cells. XII. A versatile system for the large-scale batch or continuous culture of plant cell suspensions. J. exp. Bot. *22* (70) 177 (1971)
58. Withers, L. A.: Studies on the growth in culture of plant cells. XXI. Fine structural features of *Acer pseudoplatanus* L. cells responding to 2,4-Dichlorophenoxyacetic acid withdrawal in turbidostat culture. J. exp. Bot. *27* (100), 1073 (1976)
59. Yeoman, M. M., Macleod, A. J.: Tissue (callus) cultures — techniques. In: Plant Tissue and Cell Culture. Street (Ed.), p. 31. Oxford: Blackwell Scientific 1977
60. Yasuda, S., Satoh, K., Ishii, T., Furuya, T.: Studies on the cultural conditions of plant cell suspension culture. Proc. IV, IFS Ferment. Technol. Today, p. 697 (1972)

61. Young, M.: Studies on the growth in culture of plant cells. XVI. Nitrogen assimilation during nitrogen limited growth of *Acer pseudoplatanus* L. cells in chemostat culture. J. exp. Bot. *24* (83), 1172 (1973)
62. Zenk, M. H., El Shagi, H., Schulte, U.: Anthraquinone production by cell suspension cultures of *Morinda citrifolia*. Planta med. (Suppl.), p. 79 (1975)
63. Zenk, M. H.: The impact of plant cell culture on industry. In: Frontiers of Plant Tissue Culture. Thorpe (Ed.), p. 1. Pub. Int. Ass. Plant Tissue Culture 1978

Embryogenesis in *Citrus* Tissue Cultures

P. Spiegel-Roy, J. Kochba
Division of Fruit Breeding and Genetics, Institute of Horticulture,
A.R.O. The Volcani Center, Bet Dagan, P.O. Box 6. Israel

The morphogenetic potential of *Citrus* tissue *in vitro* has been realized to full extent in explants from the somatic cells of the nucellus. The role of apomixis and polyembryony in *Citrus* evolution, propagation and breeding has been evaluated. While nucellar tissue has given rise to embryos *in vitro* in polyembryonic, and in rare cases, monoembryonic cultivars, it also produces an embryogenic callus in many *Citrus* species and cultivars. Conditions and factors for realizing its embryogenetic potential have been described. Lowering the auxin status, anti-auxins, irradiation and the use of galactose-yielding sugars have enhanced embryogenesis. Protoplast cultures have been established from the nucellar callus, also giving rise to plants via newly formed callus. Callus has maintained stability under prolonged subculture and remained diploid. Selection of callus lines tolerant to diverse stresses (salinity, 2,4-D) has been initiated and stress-tolerant, embryogenic callus lines have been established. The significance of these culture methods for mutation breeding is discussed.

1 Role of Polyembryony and Apomixis in *Citrus* Propagation and Breeding

The genus *Citrus* and its wild relatives are members of the family *Rutaceae*, subfamily *Aurantioideae*[147)]. The importance of *Citrus* to agriculture and the economy is demonstrated by its large-scale production, commerce and world wide distribution. All species of the genus *Citrus* are evergreen, but the related genus *Poncirus* is deciduous. A large variability exists within the genus, and closely related genera, often interfertile, provide an even wider array of characters[159)]. However, there are important barriers to the full utilization of this variability, one of which is the apomixis exhibited by a great number of species and varieties. Diploidy is the general rule in *Citrus* as well as related genera, the somatic (2n) chromosome number being 18[43)].

In many varieties of *Citrus* and in the related genera, *Poncirus* and *Fortunella*, additional embryos, derived not from cells of the embryo sac, but by mitotic division from somatic cells of the nucellus develop in the ovule[49)]. Before the first egg undergoes division, a few large cells, with large nuclei and dense cytoplasm are found in the nucellus[43)]. These cells are located mainly near the micropylar end of the embryo sac. Some of the cells divide, with the new cell mass protruding into the embryo sac, in vicinity of the embryo derived from the egg cell. Fertilization of the egg cell in citrus has been reported to occur from 3 days to 4 weeks after pollination. Cell division of the zygote starts soon thereafter. The endosperm is already multicellular at this stage, partly filling the embryo sac and disappearing at a later stage. The extra embryos derived from the nucellus are called nucellar or adventive embryos in contrast to zygotic or sexual embryos. They have a rather irregular shape, and may or may not posses a suspensor. The relationship of the initiation of nucellar embryos with the process of pollination and fertilization is not clear. It is probable that pollination is usually, if not invariably necessary. There is evidence that with certain interspecific hybrids, pollination is definitely required for the formation of nucellar embryos[43)]. Somewhat less clear is the need for fertilization. Some self-sterile varieties occasionally produce fruits with imperfect seeds and underdeveloped embryos after self-pollination or incompatible cross-pollination. The embryos that arise may be of nucellar origin. Usually, however, nucellar embryos develop after fertilization.

In a little known *Citrus* relative, *Aegle marmelos* (L.) Corr. the egg and antipodal cells degenerate, while both endosperm and nucellar embryos may develop normally[64)]. This is a case of nucellar embryos developing without the initiation of a sexual embryo. Possibly a fusion between polar nuclei and a sperm cell did occur. As no male cells have contributed to the formation of nucellar embryos and no reduction division has occurred in seed parent cells giving rise to them, the nucellar embryos are considered identical in genetic constitution with the seed parent, except for differences that may have arisen as a result of somatic variation. This type of apomixis has important consequences for evolution, breeding and culture of *Citrus* fruits as well as for tissue culture practices and considerations.

A very high degree of heterozygosis exists in *Citrus* due to gene mutation, ease of cross-pollination and prevalence of nucellar embryony. Seed reproduction by nucellar embryony preserves any heterozygosis that may have originated either by hybridization or mutation. This process favors the accumulation of mutant recessive genes, including genes detrimental to fertility. Successful sexual reproduction, especially by selfing is thus at a disadvantage. Through the process of natural selection recessive genes accumulated are being preserved because of nucellar embryony. The latter has also acted as an isolating agent favoring evolutionary differentiation[27]. We can thus surmise that many, but not all forms of *Citrus* and their progenitors have reproduced since antiquity entirely by nucellar embryony. An important differentiation of varieties has occurred mainly via somatic changes. Thus, while mandarins (*C. reticulata* Blanco) have a broader genetic base and have been further differentiated also by sexual reproduction, we find a rather low variability, corresponding essentially to somatic variation among sweet orange, (*C. sinensis* L.), grapefruit (*C. paradisi* Macf.) and true lemons (*C. limon* L.) On the other hand, all tested varieties of citron (*C. medica* L.) and pummelo (*C. grandis* L.) are monoembryonic and produce only zygotic seedlings.

Within the genus *Citrus*, nucellar embryony is inherited in a rather simple fashion[115, 62]. Hybrids from sexual, monoembryonic parents are essentially monoembryonic. Nucellar embryony is thus considered to be principally controlled by one dominant gene, with monoembryonic varieties being homozygous and recessive. However, some crosses between certain monoembryonic *Citrus* and the polyembryonic *Poncirus* have been found to yield only monoembryonic progeny[110], a fact that points to a more complicated gene interaction.

In all genotypes in which nucellar embryos are produced, the zygotic embryo has to compete for survival with one, and generally more, nucellar embryos. The lower the number of embryos per seed, the larger the average embryo size and the probability for the sexual embryo to survive. From one to forty adventive embryos have been found, in different varieties, by Furusato et al.[45]. Toxopeus[155] lists 3 factors relatively unfavorable to zygotic embryos.

1) the fecundated cell starts to divide when most nucellar embryos are at least few celled;
2) the zygotic embryo, located near the apex of the embryo sac, is in a less favorable situation trophically;
3) it may be weaker in genetic constitution.

The number of embryos per seed varies according to variety, pollinator, position on tree and further, fluctuates significantly in response to environment[156]. The pollen parent may sometimes influence significantly the proportion of zygotic seedlings. When self-pollinated, rough lemon produced almost exclusively nucellar seedlings; in one experiment, when it was pollinated by *Poncirus*, 46% of the seedlings were zygotic[28]. A recent report points to gibberelin applications influencing the percentage of zygotic seedlings in a highly polyembryonic variety[36].

Many varieties, classified as monoembryonic, never produce nucellar embryos *in vivo*. However, they may, rarely, give rise to more than one embryo per seed, by fission of the zygote or through fertilization of more than one egg in a single ovule[114].

Theoretically all nucellar trees should be genetically identical to the mother tree. Phenotypic variability is in many cases likely to be due to juvenility and differences in juvenility expression. It has been suggested that pollination and especially fertilization could affect genome expression in the nucellar embryo. While Iglesias et al.[59] found peroxidase zymograms to vary between plants of the same cultivar and seedlings from the same seed, this is not borne out by results obtained in our laboratory[26,142]. The chimeral nature of certain *Citrus* cultivars might explain some of the variation found in nucellar offspring[135,15]. Also various proportions of nucellar cells may carry the mutation reflected in the phenotype of the mother tree[109].

No significant problem because of lack of uniformity in seedlings used as rootstocks seems to exist in *Citrus* because of the fact that only highly polyembryonic clones are selected for the provision of uniform seedlings.

When cross-breeding is employed in the production of new varieties, nucellar embryony proves a definite obstacle in many cases. The combination of apomixis and of considerable or complete pollen sterility is especially troublesome; such varieties cannot be used as pollen parents, and when used as seed parents only a very low proportion of hybrids is found[43]. Whenever a search is made for variation in nucellar offspring, nucellar fertility, and probably also some degree of gametic fertility are required.

For breeding, rootstock and experimental work it is vital to be able to distinguish between zygotic and nucellar offspring at an early stage. This is most difficult to accomplish, using morphological characteristics only, exception a few cases[55,141] and especially those in which the pollen parent bears distinct, dominant characters, such as the trifoliate leaf of *Poncirus trifoliata L.*[28]. However, valuable biochemical methods for distinguishing zygotic from nucellar progeny at an early stage have been developed lately. These include flavonoids[149,150], peroxidase isozymes[42,142], oxidative browning due to polyphenols[41] and also long-chain hydrocarbons[111].

Since most *Citrus* viruses are seldom seed-borne, nucellar seedling budlines have given rise to satisfactory, vigorous stionic (rootstock-scion) combinations which would otherwise fail or show decline when virus infected scion buds are used[27]. Juvenility in such nucellar selections is rather prolonged and evidenced by thorniness, vigorous, upright growth habit, slowness in reaching bearing age, alternate bearing habit and deviating fruit characteristics. However, their use and the essentially nucellar combination (both in rootstock and scion) has often resulted in longer-lived and more prolific trees. Lately, a much higher incidence of *Phytophtora* crown and root rot has been found to be associated with the use of certain nucellar scions budded on *Phytophtora*-sensitive rootstocks, when compared to the performance of old budlines on the same rootstock. (Unpublished results). The use of nucellars as a source of virus-free material in established cultivars will probably loose in importance because of the more recently developed methods of thermotherapy and of shoot-tip grafting *in vitro*[65,100,105,106]. The latter methods have enabled the obtaining of virus-free material without juvenile characteristics and an attainment of earlier bearing compared to the use of nucellar scions.

2 Morphogenesis in *Citrus* in vitro

2.1 The Culture of Ovules and Seeds

The culture of developing seeds at various periods after the fertilization of the ovule was initiated as early as 1958 by Maheshwari and Rangaswamy[87]. Further studies with fertilized ovules and seeds were made by Rangaswamy[123,124] and Sabharwal[130].

The cultured explants, containing partially or fully developed nucellar and possibly zygotic embryos, gave rise to additional embryos as well as embryo-forming callus. Only in a few cases were new plants established.

Although Maheshwari and Rangaswamy[87] pointed out the possible significance of ovule culture for breeding, this line of study was not continued. It is of interest to note that at the same time, 1958, Melchers and Bergman[94] described the potentials of plant tissue cultures for the isolation of mutants. One can assume that the complex origin of the tissues arising from cultured developing seeds, uncertainty as to the origin of the additional embryos and difficulties in rising plants from them limited further interest in this line of work.

The successful culture of mature *Citrus* embryos, which were probably of nucellar origin was first reported by Ohta and Furusato[112]. A decade later Rangan et al.[121, 122] induced the development of nucellar embryos in nucellus explants of naturally monoembryonic *Citrus* cultivars and promoted the further development of heart-shaped embryos of zygotic (sexual) origin from naturally polyembryonic cultivars. Their aim was to obtain virus-free true to type plants from monoembryonic cultivars as well as to help survival of hybrid plants from progenies of crosses involving polyembryonic female parents. Other methods developed, such as micropropagation to obtain virus-free plants[105, 106] and the development of criteria for the early detection of hybrid seedlings[26, 142] have since reduced interest in this approach.

The success in culture of tissue and in establishment of plants was largely due to the use of the Murashige and Skoog (MS) culture medium[101] which not only is more concentrated than the White medium[163], but contains more N, a considerable proportion of it as NH_4^+. Modifications to MS medium for *Citrus* by Murashige and Tucker[102], namely a considerable increase in vitamins, further improved the growth of *Citrus* tissues in culture.

While Maheshwari and Rangaswamy[87] assumed that only fertilized ovules respond favorably to culture conditions, it was later proved by Bitters et al.[16] as well as by Button and Bornman[21] that nucellar embryos could be obtained also from nucellar explants extracted from unfertilized ovules. We[70, 81] and Mitra and Chaturvedi[98] showed that in addition to embryos also a nucellar embryogenic callus can be obtained from unfertilized ovules and nucelli. Unpollinated ovaries also yielded embryos. Mitra and Chaturvedi[98] observed first a decline and later a cessation in embryogenesis under more prolonged subculture. We obtained an embryogenic callus from unfertilized ovules and nucelli of the Shamouti orange *Citrus sinensis* L. Osbeck in which specific conditions for the maintenance of embryogenesis were established[67]. This callus as well as callus obtained later on from other Citrus cultivars (unpublished results) form the basis for morphogenic and breeding studies which will be described in greater detail later on.

2.2 Morphogenesis from Other Tissues

There is only one report on shoot morphogenesis obtained from cultured tissues other than ovules and seeds. Adventitious buds were induced in callus cultures derived from the stem tissue of seedlings (*C. madurensis* L. cv. Calamondin) by Grinblat[51]. Under suitable conditions these buds formed shoots and could be rooted to yield established plants. Although it was claimed that buds differentiated from the callus tissue the evidence given is not decisive. It is possible that the buds originate from the stem explants outgrowing the callus which covered the stem sections in culture. No attempt was made to obtain morphogenesis from subcultured callus or isolated cells as the intention of the author was to obtain true to type virus-free plants.

Seedlings of *Citrus* have a tendency to form adventitious buds when they are cut back just above the cotyledons and when kept in a humid environment[60, 66, 131, 137]. The ability of decapitated seedlings to differentiate buds was applied in a mutation breeding program for *Citrus*[66, 137, 138, 139].

We have observed root formation in subcultured callus from lemon (*C. lemon* cv. Villafranca) anther filaments when these were maintained on a kinetin: IAA grid. Root-formations was sporadic and was not accompanied by bud formation (unpublished results).

The culture of isolated buds was carried out by Altman and Goren[2, 3, 4, 5]. The aim of these studies was to elucidate the role of growth regulators and environmental conditions in bud development. Although work was done with meristems and not single cells or undifferentiated tissues, interesting morphogenetic effects were observed. GA_3 promoted internode elongation and cytokinins stimulated adventitious bud formation. IAA delayed and ABA inhibited sprouting of buds. A relatively high sucrose level was needed for shoot growth. ABA was found to stimulate callus formation in the abscission zone of the bud. GA_3 alone was only slightly active but manifested a synergistic effect on callus growth when applied in combination with ABA.

3 Factors Affecting Embryogenesis in vitro in Autotrophic *Citrus* Cultures

3.1 Characteristics of the Subcultured Nucellar *Citrus* Callus

3.1.1 Establishment of embryogenic Cultures

Work on culture of *Citrus* ovules in our laboratory started in 1971 with the primary object of inducing development of adventive embryos in nucellar explants from unpollinated ovules, thus securing solid mutant development following mutagenic treatments. Although embryos and plants were obtained[81] for several *Citrus* cultivars their frequency was low. Extraction of nucelli in most cultivars is tedious, rendering large-scale work for breeding studies rather impractical. On the other hand, callus developed from explants of *C. sinensis* cv. Shamouti. Subculturing, under suitable conditions gave rise to rapid proliferation and the formation of numerous green embryos[70]. After several subculture periods embryogenesis declined. The basic medium (MS) devoid of any growth substances which for several passages failed to support growth was now able to support excellent growth. This indicated habituation (autotrophy) of the callus with regard to auxin and cytokinin requirements. Phytohormone autotrophy is a regular phenomenon in *Citrus* cultures of nucellar origin and has been observed so far also in cultures established from nucelli of other *Citrus* species. Callus has since then (7 years) been subcultured and maintained on basic medium only. Transfer of callus after 10–15 weeks in culture greatly increased embryogenesis while transfer after 4–5 weeks considerably reduced embryogenesis[67].

After repeated subculturing callus lines were selected differing greatly in embryogenic potential, ranging from abundant to almost none. These callus lines served for the study of factors affecting embryogenesis.

3.1.2 Anatomical and Development Studies

The origin of embryos has a great bearing on the use of *Citrus* tissue culture for mutation studies. Embryos have to be derived from cells and not from meristematic bud-like centers to permit the development of solid mutants after mutagenic treatments. Light and electron microscopic studies[24, 25)] have indicated that embryogenesis can be traced back to single cells of the nucellar callus. A comparison of embryogenic and non-embryogenic lines showed that in all of them development is in the form of proliferation of globular bodies which reach a larger size in non-embryogenic lines as well as in embryogenic lines, under non-inducive conditions for embryogenesis. Their growth (callus proliferation) continues by budding off of new globular bodies from existing ones. Under embryogenic conditions the globular bodies are smaller and cells on the periphery as well as within them form embryos (unpublished results).

3.1.3 *Citrus* Protoplasts

Isolation and culture of protoplasts is an important tool for genetic studies with cell cultures. It permits cloning from single cells and is a prerequisite for studies aimed at inducing genetic modifications by introducing foreign DNA or cell particles. The culture of protoplasts also opens the possibility for the study of interspecific or intraspecific fusion hybrids[10, 47, 56, 116, 117, 118)].

Protoplasts have been isolated from many species and sustained divisions leading to colony formation have been obtained in some of them[10, 161)]. Recovery of whole plants from protoplast cultures have been reported for 14 species, fruit tree species being absent[10)]. It is therefore of great interest that Vardi et al.[160)] were successful in isolation and culture of protoplasts from ovular callus of the 'Shamouti' orange. The morphogenetic capacity of this callus was not impaired. Thus the full sequence callus→protoplast→cloned callus colony→plant was established for the first time for a fruit tree species[158, 160)].

Sustained divisions in isolated protoplasts are dependent on a rather high minimum cell density[104)]. For orange protoplasts the optimal density was found to be 10^5 ml^{-1}, and 10^4 ml^{-1} proved to be the minimum cell density for colony formation. The high optimal plating density required often leads to fusion of adjacent colonies and reduces growth of individual colonies before they can be effectively diluted[158, 161)]. Protoplast density was successfully reduced by using the X-ray irradiated feeder protoplast method for tobacco described by Raveh et al.[125, 126)]. Utilizing the ability of orange protoplasts to grow in absence of auxin in the medium (tobacco protoplasts are auxin-dependent) or using irradiated orange feeder protoplasts, the method was modified for the culture of orange protoplasts by Vardi and Raveh[161)]. Optimal plating density was thus reduced to 10^3 ml^{-1}.

Mutation studies with protoplasts are now performed and the isolation of protoplasts from additional *Citrus* species is attempted.

3.2 The Role of Auxins

3.2.1 Auxins and Inhibitors

Several growth-substances and culture-conditions were found to either stimulate or suppress embryogenesis in *Citrus* callus[67, 70, 72, 73, 74, 75, 76, 78, 79, 80, 136, 153, 154]. Although some or all of these factors could exert their influence independently a special role of auxin seems strongly indicated. The autonomous growth habit of the 'Shamouti' callus in our experiments indicates an adequate endogenous level of auxin and possibly also of other growth-substances. We assume that manipulating the endogenous auxin level can lead to suppression of embryogenesis by raising the level and a stimulation can be achieved by lowering it. The assumption that auxin is a major factor controlling embryogenesis is supported by a substantial number of reports[6, 52, 92, 107, 127, 144, 151] on other plant tissues in culture, most of them auxin-dependent for their growth. The autonomous growth habit and the availability of lines differing in embryogenic capacity make *Citrus* callus suitable for the study of the effects of auxins on embryogenesis.

We found that the addition of auxins (IAA, NAA and 2,4-D) to the culture medium significantly or completely suppressed embryogenesis while inhibitors of auxins synthesis (7-Azi-indole and 5-HNB) strongly stimulated this process. These results were obtained with callus lines showing an embryogenic potential. When non-embryogenic lines were subjected to the same treatments, embryogenesis was not induced. These lines were not significantly affected in their growth by high auxin levels causing a significant reduction in growth of cultures from embryogenic lines[75].

Aging of the callus by increasing the interval between subcultures from the standard 5 weeks to 10–15 weeks also stimulated embryogenesis. This was our first approach to obtain embryogenesis which otherwise declined[67]. Auxin content was reported to decline in aging tissues[46], thus indicating furthermore a possible morphogenic effect of auxin.

3.2.2 Peroxidase and IAA-Oxidase

Peroxidase and IAA-oxidase activity as well as changes in isoenzyme pattern may be considered part of the normal aging process[83, 133]. The *in vivo* significance of these enzymes in the regulation of endogenous auxin level is rather disputed[119, 120, 132]. However, a possible morphogenic role for these enzymes was suggested[46, 83, 157, 165].

We found that peroxidase and IAA-oxidase activities were significantly higher in embryogenic compared to non-embryogenic lines[69]. Treatments inhibiting embryogenesis reduce the enzyme activity and vice versa. Differences in isoenzyme patterns were also observed when embryogenic and non-embryogenic lines were compared. Although it is not fully established whether increased activity is the cause or result of embryo formation we have some evidence that changes in enzyme activity precede embryo formation (unpublished results).

3.2.3 IAA Metabolism

Although we have not yet studied the actual endogeneous IAA level in *Citrus* callus we studied the uptake and metabolism of IAA, using IAA-1-^{14} and IAA-2-^{14}C. We established that the embryogenic callus converted a great proportion of IAA very rapidly into IAA-aspartate[38] which is considered a very stable conjugate by which excess IAA is removed from tissues[35]. A non-embryogenic line formed very little IAA-aspartate and this only after much longer incubation periods. The ability to form this conjugate possibly reduces auxin to a level more compatible with embryogenesis.

3.3 The Effect of Irradiation

Exposure of the callus to gamma irradiation resulted in a marked stimulation of embryogenesis. The maximal stimulation was obtained at 12–16 kR. Lethality was observed at 28–32 kR. When experiments were carried out with non-habituated callus and the medium thus contained auxins it was possible to mediate the radiation effect by irradiating the medium alone[136]. In later studies with habituated callus, growing on a medium devoid of auxin, no such medium effect was observed, while callus irradiation continued to stimulate embryogenesis[74].

Radiation treatments have been reported to lower auxin levels[32]. We assume that this effect could provide the explanation for the irradiation-induced stimulation of embryogenesis. On the other hand, one must consider that subcultured explants contain various stages of development, from single cells to globular bodies and already initiated embryos of various sizes. Single cells and smaller aggregates can be expected to be more sensitive to irradiation thereby relieving more developed and larger embryonic stages from competition and resulting in the appearance of more embryos. This could at least explain the increased variation in response of individual cultures with progressing age of callus used for subculture[74].

In recent experiments in which isolated protoplasts were irradiated, highly embryogenic colonies were obtained[158]. This favors the assumption of a direct effect on cells. Furthermore, the addition of non-irradiated IAA to the irradiated callus resulted in a dose-dependent phenomenon. While at low radiation doses (less than 8 kR) embryogenesis was inhibited by IAA, the inhibition was gradually overcome by doses higher than 8 kR. Some radiation doses + IAA levels (12–16 kR) even stimulated embryogenesis and improved the rate of development of the embryos. Usually lethal radiation doses (30 kR) still enabled survival and growth of cultures in the presence of IAA[74]. We postulate that with certain radiation doses auxin is reduced to suboptimal levels. Added IAA would then promote embryogenesis, as was indeed observed.

3.4 The Effect of Other Plant Growth Regulators

Embryogenesis was suppressed following additions of cytokinins. In the presence of kinetin and IPA inhibition increased with concentration while the response to BA fluctuated. 8-Aza-guanine (AG) clearly stimulated embryogenesis even at very low (0.001 mg l^{-1}) concentrations[75]. Although AG has been reported to have antagonistic effects on cytokinin activity[17], it is also known to be a mutagen and probably has other metabolic effects[85].

Gibberellic acid (GA_3) suppressed embryogenesis strongly when applied at 1 to 30 mg l^{-1} and completely between $50\text{–}100 \text{ mg l}^{-1}$. Two inhibitors of GA_3 biosynthesis, 2-chloroethyl trimethylammonium chloride (CCC) and succinic acid, 2,2-methylhydrazide (Alar) significantly stimulated embryogenesis at concentrations up to 10 mg l^{-1}.

Although ABA is classified as an inhibitor, an increasing number of reports indicate some growth stimulating responses as well as morphogenetic effects of exogenously supplied ABA in plant tissue and cell cultures[6, 12, 44, 61, 84, 97)]. A marked increase in embryogenesis was obtained by low ABA levels (0.04–4.0 μM). Higher concentrations, 40 μM–80 μM drastically suppressed embryogenesis but were found to support growth[76)]. Similar results were obtained by Tisserat and Murashige[153)] who tested rather high concentrations.

Because of the apparent relationship between auxin activity and ethylene evolution, we also examined the effect of ethylene, supplied as ethephon (2-choloroethyl-phosphonic acid). Low concentrations of ethephon (0.01–1.0 mg l^{-1}) significantly stimulated embryogenesis[76)]. Concentrations higher than 1 mg l^{-1} suppressed embryogenesis. The suppression of embryogenesis in *Citrus* and carrot cultures by higher ethephon levels was also observed by Tisserat and Murashige[154)].

3.5 The Effect of Sugars

We had observed that omission of sucrose from the medium for a single culture passage, although resulting in complete cessation of growth, greatly stimulated embryogenesis when the following subculture was performed on a sucrose-containing medium[67)]. A report by Anker[7)] on the inhibition of auxin biosynthesis by galactose, lactose and raffinose led us to study the effect of these sugars on embryogenesis. Galactose, lactose and raffinose when substituted for sucrose in the culture medium stimulated embryogenesis very considerably[79, 80)]. All these test sugars also supported very good growth in absence of sucrose. Non-embryogenic callus lines that had not responded to other embryo-stimulating treatments were for the first time induced to form embryos on rather high galactose levels (6–10 percent). Very low galactose concentrations, unable to support growth, were also effective in stimulating embryogenesis in the embryonic lines.

Addition of IAA and NAA over a concentration range of 0.01–5.0 mg l^{-1} progressively inhibited embryogenesis in the presence of 1–5% galactose or lactose. The magnitude of inhibition caused by a given auxin concentration was not reduced by increased galactose or lactose levels. This indicates that the mode of action of these sugars cannot be ascribed to exogenous auxin inactivation[79, 80)].

Both the observed effects of galactose and galactose yielding sugars — support of good growth and stimulation of embryogenesis — are of great interest. Very few plant tissues grow on galactose[57] and this ability might be utilized in protoplast fusion studies.

When the standard concentration of 5 percent sucrose was added to the medium, the stimulating effect of galactose, lactose or raffinose was entirely suppressed. In studies with ryegrass (*Lolium multiflorum* Lam.) it was found that sucrose and glucose counteract the inhibition of root elongation caused by galactose. It was also established that glucose markedly reduced uptake of galactose by roots[88)]. Sucrose may have had the same effect. If galactose inhibits auxin synthesis[7)] counteraction of sucrose could be due to stimulation or mobilization of auxin[63)]. Still, Anker[7, 8)] did not observe that the galactose inhibition of auxin synthesis in Avena coleoptiles was antagonized by other sugars. We are now studying the effects of several sugars and sugar combinations on growth and embryogenesis.

Galactose is considered non-toxic and growth-supporting to only very few tissues in culture[86, 108, 128, 143)]. The toxic effect for most tissues is due to their inability to convert galactose-1-phosphate and or UDP galactose into UDP glucose[58, 91, 128)]. Although our results seem to support Anker's[7)] findings, we have no direct evidence yet for an inhibition of auxin synthesis in *Citrus* callus by galactose. Galactose was shown also to reduce auxin movement[82)] and to inhibit synthesis of cell wall components[58, 113)] due to galactose-1-phosphate accumulation in tissues for which galactose is toxic[49)].

Of much interest is a report providing evidence that galactose stimulates ethylene evolution[33)]. High levels of ethylene have been found to suppress auxin biosynthesis[82)]. It is therefore possible that galactose regulates auxin biosynthesis via ethylene. Moreover, we have obtained a stimulating or suppressing effect on embryogenesis with ethephon, depending whether a low or high (less or more than 1 mg l^{-1}) concentration was used[76)].

In more recent experiments (unpublished results) embryogenesis was inhibited in tightly sealed culture vessels. Such vessels contained high levels of ethylene in a range that had been found to inhibit embryogenesis[76, 154)].

Though many of our findings as well as those of Tisserat and Murashige[153, 154)] point to auxin as the decisive factor in embryogenesis in *Citrus* one has to assume that the probably complex processes leading to embryogenesis would be under the control of more than one hormone. Interactions with auxin have been reported not only for ethylene but also for ABA or GA_3[8, 96, 162)].

Further studies will be carried out in order to gain additional information on the complex events leading to embryogenesis. We hope that the information will be also helpful in efforts to induce embryogenesis in other plant tissue cultures.

3.6 The Effect of Vitamins and Organic Substances

Optimal levels for proliferation of citrus tissues of several vitamins were established by Murashige and Tucker[102)]. There are few reports on morphogenetic effects of vitamins. Pseudodulbil (globular embryo) formation directly from nucellar tissue of *Citrus sinensis* cv. Washington Navel was reported by Button and Bornman[21)] to be stimulated by ascorbic acid. Ascorbic acid was also alleged to enhance growth of *Citrus Natsudaidai* plantlets[112)]. We observed that certain ascorbic acid concentrations (1 g l^{-1} to 10 g l^{-1}) were toxic for habituated *Citrus* ovular callus. Different levels and combinations of inositol, pyridoxine HCl, nicotinic acid and thiamine HCl had no effect on embryogenesis in the same callus (unpublished results).

Complex organic substances of natural origin are usually of incompletely defined composition[23)] and their use should whenever possible be avoided in systematic studies. However, malt extract (ME) was reported to be beneficial to embryogenesis in nucellus cultures[21, 81, 98, 121, 122)], as in development of pseudobulbils from single cells[22)]. Subcultured habituated nucellar *Citrus* callus rapidly lost dependence on ME for embryogenesis[70)].

Coconut milk (CM) has been widely used in tissue culture[145)]. Similarly to ME there are several reports on the beneficial effects on embryogenesis in nucellus cultures. Its effect on subcultured and habituated callus was not studied, because other, well-defined substances have been found to exert a potent effect on embryogenesis (see former chapter).

Yeast extract (YE) and casein hydrolysate (CH) seem to be ineffective in stimulating embryogenesis. However, YE improved organ development in cultured nucelli[124) 134)], and CH, separately or combined with YE promoted the maturation of immature embryos and their further development into plantlets[134, 152)]. Orange juice appears to stimulate growth of *Citrus* callus from several tissues in some cases[20, 39)] but not in others[102)]. No effect on embryogenesis in habituated citrus callus was observed (unpublished results).

It appears in many cases that to obtain similar effects organic substances could be substituted by well defined compounds, such as growth hormones and others[70, 98)]. However, because of our incomplete knowledge, the use of organic substances might be beneficial in inducing or promoting several stages of embryo and plant development, under cartain conditions.

3.7 Totipotency in *Citrus*

Although totipotency in *Citrus* nucellar callus has been proven, it is not yet clear whether all cells are totipotent or whether totipotent cells are distributed among non-totipotent cells. The latter may or may not have an assisting function as to embryogenesis. Usually only a small proportion of cells present in the solid culture medium develop into embryos. This could be due to competition for limited space and nutrient:. It was observed that in suspension cultures under suitable conditions most incubated cells turned into embryos (unpublished results) yielding approximately 100× more embryos than inocula of the same size on agar medium.

Recent work with Shamouti protoplasts[158, 160, 161)] from which embryos were recovered from virtually every single protoplast-derived colony[158)] supplies further evidence as to the totipotency of the cells. This is, however, no clear indication of the totipotency of all cells as only a small proportion of cells form viable protoplasts and only a small fraction of the isolated protoplasts developed colonies. This could be a matter of techniques but also a preferential isolation as well as colony formation from totipotent cells. Furthermore, colony formation (however small) seems to be a prerequisite for the initiation of embryo development. The direct development protoplast→embryo without an intervening callus phase has not yet been clearly demonstrated.

Whether or not all cells of the citrus callus are totipotent or not, this system can be applied for the induction and selection of mutants because cells and protoplasts have the potential to regenerate plants. It would only mean that we are not dealing with "millions" but only a "few thousands" of potential plants per culture dish, — still a tremendous advantage over the field approach in mutation induction.

The nucellus which gave rise to the subcultured callus has the capacity to differentiate embryos *in vivo*. It is therefore indicated that embryogenic potential is an inherent property of cells in culture originating from the nucellus. Embryos are differentiated from the nucellus *in vivo* preferentially near the micropylar region while usually few or no embryos are formed from cells close to the chalaza[40, 43)]. We may assume that our non-embryogenic lines originate from cells near the chalaza region. These could be readily selected on subculture because no embryos are formed by them. Embryogenic lines would preferentially originate from nucellus cells adjacent to the micropyle, and might, because of difficulties in stringent selection, still contain a certain proportion of cells deriving from the chalazal region. Comparative studies on embryo induction in embryogenic and non-embryogenic lines could provide further information on behavior of cells in which embryogenic potential *in vivo* is suppressed.

The subcultured nucellar *Citrus* callus has maintained the embryogenic potential manifested in the original cells of the nucellus. Certain culture conditions could suppress this ability, others resuppress it. This basic assumption must be kept in mind when evaluating the broader significance of factors which were found to affect embryogenesis in our subcultural citrus callus. While of great potential importance for the use of *Citrus* cultures for breeding studies, they might also hopefully contribute to the understanding of morphogenesis in comparable systems, i.e. ovular or nucellar tissues of other species in culture. There is no evidence as yet that results from *Citrus* can be applied to cultures from other plant species or tissues of totally different origin the totipotency of which might be genetically blocked.

4 Development of Embryos and Establishment of Plants

The establishment of plants from tissue and cell cultures is the final step for the utilization of these methods for plant breeding or propagation purposes. Great progress has been made in the latter field. The experience gathered has been summarized by Murashige[99].

Embryos obtained from *Citrus* cultures are in various stages of development at a given time. Four stages were defined so far[68]. Stage 1 embryos are globular and those of stage 2 elongated. These have neither root or shoot primordia. More advanced stages have cotyledons as well as root primordia in stage 3. Shoot primordia are found in stage 4.

Rooting was promoted in all stages of embryonal development. GA_3 alone or in combination with adenine sulfate (ADS) was highly root-promoting while hormones usually known as stimulants of rooting — NAA or IBA — suppressed rooting. Stage 1 and 2 embryos did not develop into plantlets after rooting. A root meristem was initiated by the treatments described but shoot meristems did not develop.

Stage 3 and 4 embryos rooted at highest frequencies and many of them also developed shoots. Most of these embryos have 3 and more cotyledon-like structures. Asymmetric development can lead to formation of misshapen embryos. After 1–2 months of culture many of these give rise to normal shoots[68, 71].

A rigorous schedule of hardening procedures is necessary in order to establish healthy plants in the greenhouse. Roots formed in agar do not function properly. We are studying substitutes for agar and try to shorten the period of culture in the agar medium. Growth of plants is slow especially from 'Shamouti' orange callus. These plants are also comparatively weak *in vivo* on their own roots. Tip grafting of the embryo shoot onto young seedling rootstocks is a promising approach and speeds up development of plants.

We have established the full sequence of development from callus cells to plant[70] and Vardi et al.[160] have established this sequence by beginning with protoplasts. A mass production of plants is not yet feasible but can probably be worked out, if necessary. Once a promising callus colony has been isolated, one plant obtained from it would be all that is required as more can be produced later on by budding on rootstocks. Nevertheless we study the factors promoting development of immature into mature embryos and intend to improve conditions for establishment and rapid growth of many plants from culture.

5 Mutant Selection in *Citrus* Callus

5.1 Significance of Tissue and Cell Cultures for Selection and Breeding

The advantages of tissue and cell cultures for selection of mutants have been extensively reviewed[18, 30, 89, 93, 94, 95, 140, 166]. The manipulation of a large number of cells/genotypes in a limited space is attractive compared with the large demands on field space with conventional breeding techniques. The realization of the potential value of cell cultures for mutant selection is, however, faced with considerable difficulties. Effective screening methods for the early detection and isolation of desirable mutants have still to be developed[18, 89, 90]. In some plants this was successfully achieved by culturing cells on growth inhibitors such as salts, herbicides, antibiotics, amino acids and nucleotide analogs. Special methods such as BUdR enrichment were also designed[11, 14, 29, 37, 48, 103, 164]. Resistant cell lines from various sources were recently reported but only in few cases has the resistant trait been studied in regenerated plants. In other cases the selected lines did not possess or may have lost the potential for regeneration during subculture[89]. Improvement of methods for regeneration of plants from cultured cell is therefore of utmost importance.

In the present stage of knowledge, selections must be made of characters that can be expressed at the cellular level. These are mostly resistance phenomena, as well as autotrophic and auxotrophic mutants[164]. Many resistant reactions (to diseases and herbicides) often depend on the structure of a tissue or an entire organ or an interaction between organs[18]. Despite these difficulties it is encouraging that mutant plants have been recovered from cell lines and progress has been made in this field. It has been suggested that to overcome the basic problems associated with cell genetics, induction of mutations, isolation and recovery of mutants, research should be concentrated on a few model species. *Nicotiana* spp., *Daucus carota*, *Datura inoxia* and especially *Arabidopsis* have been claimed as most suitable[89]. Conditions for callus and protoplast cultures have been established for these species, and plants can be regenerated and analysed genetically. Although this may be correct for studies designed to solve basic problems, cell genetic studies on plants of economic importance are as well justified.

Citrus ovular callus offers many of the properties necessary for mutant selection. Embryos originate from single cells, the capacity for embryogenesis is maintained seemingly unlimited during subculture, probably because cells remain usually diploid. Plants can be also regenerated from cloned colonies developing from protoplasts[158].

So far no haploid *Citrus* cultures are available. Their advantage in genetical studies, in the recovery of recessive mutants and establishment of homozygous plant as a basis for planned breeding is clear. On the other hand, for the purpose of crop selection, the use of haploids in *Citrus* and other highly heterozygous species would be of limited value. Starting from a certain cultivar the chances of obtaining the same genotype after rediploidization would be minimal[50].

Genetic analysis of F_1 und F_2 populations from regenerated fertile *Citrus* plants would obviously be difficult due to the prolonged juvenile phase and because of polyembryony. But, once a desirable and stable variant plant is regenerated, whether it represents a true mutation or an epigenetic change in gene expression, it can be maintained and multiplied by clonal propagation.

We are at present studying the possibility of obtaining resistance or increased tolerance to sodium chloride and 2,4-D in *Citrus* callus.

5.2 Evaluation of Selection Techniques

Selection have so far been carried out with callus explants only. The advantages of this method are simplicity and in the relative intensitivity of calluses to possible side effects of the selective agent, such as increased molarity of the salt-containing medium. Explants contain approximately 1×10^5 cells. Subcultured at 5 week intervals the number of cell increases ten to thirty times and they constitute a loosely packed mass of globular structures. From these the embryos arise. Not all cells are in direct contact with the selective medium and gradients of the selective agent can be expected to exist throughout the callus mass. This, as well as crossfeeding between cells, might also enable susceptible cells to maintain divisions for some time. Therefore, a series of selection passages was devised, which hopefully would result in a substantial increase in amount of resistant cells in the selected cultures.

The disadvantages of prolonged selection (not crucial for *Citrus* as embryogenic capacity is maintained) might be overcome by applying selection pressure to isolated protoplasts. Colonies formed from surviving protoplasts can be further tested for resistance. The realization of this advantage is complicated if one considers the expected low frequency of mutants, 1×10^{-8} to 1×10^{-6} cells[146)] resulting in very low plating densities below the range necessary for plating even by using the feeder technique[125, 126)]. By application of a concentration of the selective agent which enables survival but no divisions, or very slowed down divisions of susceptible cells this problem might be solved. Growth of treated tissue for several cell cycles before protoplast isolation could also be considered.

Suspension culture techniques can be applied to *Citrus* but clumps of 100 and more cells are formed. The better contact with the medium offers an advantage over selection procedures used with callus explants. Further studies are under way to evaluate the relative merits of the different approaches to selection.

5.3 Selection of Callus Lines for Sodium Chloride Tolerance

In many areas of the world, irrigation water containing a higher saline concentration is available but cannot be used with many crops. *Citrus* is rather sensitive to salinity. Irrigation water containing more than 200 mg l^{-1} of Cl^- and soil salinity higher than 5 milli equivalents l^{-1} of Na^+ or Cl^- significantly reduced yield and development. Any increase of tolerance to salinity in *Citrus* is therefore of importance[31, 34, 54)].

Conventional breeding for salt tolerance is limited[148)]. In a few species large varietal differences, in many other only small differences in salt tolerance were found[53, 129)]. In soybean a dominant chloride excluder allele and a recessive includer allele determine varietal tolerance[1)]. Besides salt avoidance, the physiological nature of saline tolerance involves increased water uptake and cytoplasmic changes in protein regulated ion uptake[1,13,129]. According to the type of stress applied, whether osmotic, total salt concentration or specific ion effects, different salt tolerant mutants are possibly obtained[48]. Isolation of NaCl tolerant cell lines was reported before but in no cases were plants established from variant lines[37,103].

Citrus sinensis callus is very sensitive to NaCl. Five g l^{-1} reduce growth by 50% and 10 g l^{-1} almost completely suppress growth. By repeated selection of callus explants callus lines with increased tolerance to NaCl were isolated. Both normal and gamma-irradiated callus lines were used for selections. Greater tolerance to a given saline concentration was observed in gamma-irradiated lines[77]. When subjected to tests for stability of tolerance, however, normal, non-irradiated callus lines demonstrated greater stability in expression of tolerance, as measured by callus growth, than irradiated callus lines (unpublished results). Tolerance has been so far obtained in callus. Experiments are under way to determine tolerance in regenerated plants.

5.4 Selection of Callus Lines Resistant to 2,4-D

No breeding for 2,4-D resistance has been reported. One case of 2,4-D resistance in cell cultures was reported but no plants were regenerated[166]. A good correlation between 2,4-D sensitivity of cultured tissues and cell suspensions and sensitivity of plants was established for 11 species, most of them weeds[167,168]. *Citrus sinensis* callus is highly sensitive to 2,4-D. A 70% reduction in growth was observed in presence of 1×10^{-5} M, 2,4-D. Higher concentrations almost completely inhibit growth. Another effect of 2,4-D is the complete suppression of embryogenesis by concentrations as low as 1×10^{-7} M.

We isolated callus lines showing substantially increased tolerance to 2,4-D, growing well in presence of 5×10^{-5} M 2,4-D and some even on 1×10^{-4} M. One of these tolerant lines maintained embryogenesis in presence of 1×10^{-4} M 2,4-D, a very significant feature. Greater tolerance, as measured by growth, was obtained from irradiated cultures. Tolerant cell lines from both normal and gamma irradiated callus maintained satisfactory stability when subjected to stability tests (growth in absence of 2,4-D for 1–4 passages). The interpretation of results is complicated because of the 2,4-D dependence of some of the selected lines, which causes reduced growth in absence of 2,4-D. Two of the selected lines showed such a response while three others showed a normal response, i.e. better or equal growth in absence of selection pressure, indicating independence of growth from 2,4-D supply[77].

Experiments are under way to establish tolerance of regenerated plants. This will be feasible for the 2,4-D tolerant embryogenic line. In other tolerant lines embryogenesis is suppressed by 2,4-D and must first be regained.

6 Conclusions

Tissue culture in *Citrus* has been much advanced by the ready response of the nucellar tissue in initiating embryos *in vitro*. However, with apomixis prevalent in many *Citrus* species and cultivars, nucellar seedlings can be easily grown from seed *in vivo* in large numbers, and moreover, for experimental purposes, their nucellar nature can be confirmed by the use of suitable morphological markers. The recovery of zygotic progeny from polyembryonic seed parents, before they are crowded out by adventive embryos, could be accomplished by use of *in vitro* methods; this might prove valuable for particular cross combinations and studies. Locating the sexual embryo in the seed by position or time of appearance has been suggested, but these methods are rather cumbersome and not always dependable. While recently developed biochemical methods for the early detection of sexual progeny in polyembryonic varieties *in vivo* will be helpful in separation of zygotic and nucellar progeny in many cases, none of the methods has so far been satisfactory for the detection of sexual progeny in crosses between *C. sinensis* cultivars.

Some effort has been consecrated to the achievement of regeneration from nucelli of strictly monoembryonic cultivars, yielding only sexual progeny, such as the Clementine mandarin. In spite of the postulated presence of substances inhibiting embryo development from nucellar cells in monoembryonic varieties some of the efforts in regenerating embryos and, subsequently, plants from such cultivars have been crowned with success. However, plants of nucellar origin obtained *in vitro* from the monoembryonic Clementine mandarin (*C. reticulata*) have been found to show variation and also deviate from Clementine phenotype[65]. Some of the variation encountered, but not all of it, could be due to a chimeral situation in the Clementine mandarin. Since the development of *in vitro* shoot-tip grafting in Citrus, a more dependable method in obtaining true to type, virus-free material from established clones in general, and especially in monoembryonic cultivars has been put into effect.

Many important *Citrus* cultivars and perhaps some species also, may have evolved from somatic bud mutations. In many of these, because of heterozygosity, long juvenile period and different degrees of sterility, improvement by hybridization and subsequent selection is difficult to envisage. Mutation breeding would therefore, constitute, in such cases, a more plausible approach for improvement, especially if methods effective in overcoming chimera formation could be developed. Indeed, definite possibilities for the starting of cultures from single cells exist in *Citrus*, whether by use of callus, protoplast or, possibly, suspension culture. Because of its high regenerative capacity and comparatively ready response, nucellar material has provided the most rewarding source for culture *in vitro*. One can thus expect to be able to induce, and effect in the future, a noticeable increase in somatic variation in polyembryonic species and cultivars. However, in order to achieve further increase in variation through induced mutations, especially in monoembryonic cultivars, regeneration from internodal stem tissue or by use of adventitious bud technique from leaf tissue would be required[19]. Work carried out with leaf cuttings of *C. limon* and *C. sinensis* (unpublished results) indicate that under certain (though rare) conditions, roots and even buds: may be formed.

Attempts to obtain haploid citrus plants in culture have not been successful so far. Haploid plants would be of great importance for genetic information and inheritance studies.

As a result of tissue culture work, a cell to plant system has been now developed and a model system for selection established in *Citrus*. Biochemical markers, such as peroxidase isozymes are proving of interest not only in distinguishing different progeny but also in the detection of differences between calli. Recent work in *Citrus* has added to our understanding of some physiological and nutritional factors in the realization of the morphogenetic potential of the nucellar system *in vitro*. However, for breeding purposes, in view of the nucellar origin of cells, special efforts will be required in order to shorten the juvenile period of plants derived from culture, especially in species with protracted juvenility such as *C. sinensis*.

7 Acknowledgement

We are pleased to acknowledge the scientific cooperation of Dr. Aliza Vardi (here) and Dr. J. Button (S. Africa) and the scientific and technical assistance of Miss Shoshana Saad, Mrs. Hannah Neumann and Mr. E. Mandel. Part of the work performed and cited by us was supported by a grant from the Gesellschaft für Strahlen und Umweltforschung, GSF, Munich, West Germany, and the National Council for Research and Development, Jerusalem, Israel. Some aspects of our work were also supported by the International Atomic Energy Agency under Research Contract 127-3. We are greatly indebted to both authorities.

8 Table of Symbols

MS	Murashige and Skoog medium
IAA	3-indole acetic acid
ABA	abscissic acid
GA3	gibberellic acid
5-HNB	2-hydroxy-5-nitro benzyl bromide
NAA	α naphthyl acetic acid
2,4-D	2,4-dichlorophenoxyacetic acid
IPA	N^6-(Δ-isopentyl) adenine
AG	8-azaguanine
CCC	2-chloroethyl trimethylammonium chloride
Alar	succinic acid 2,2-methylhydrazide
Ethephon	2-chloroethyl phosphonic acid
UDP	uridine diphosphate
CM	coconut milk
YE	yeast extract
CH	casein hydrolysate
ME	malt extract

9 References

1. Abel, H. G.: Crop Science *9*, 697 (1969)
2. Altman, A., Goren, R.: Plant Physiol. *47*, 844 (1971)
3. Altman, A., Goren, R.: Physiol. Plantarum *30*, 240 (1974)
4. Altman, A., Goren, R.: Physiol. Plantarum *32*, 55 (1974)
5. Altman, A., Goren, R.: Acta Horticulturae *78*, 51 (1977)
6. Ammirato, P. V.: Bot. Gaz. *135*, 328 (1974)
7. Anker, L.: Acta Bot. Neerl. *23*, 705 (1974)
8. Anker, L.: Acta Bot. Neerl. *24*, 339 (1975)
9. Bacchi, O.: Bot. Gaz. *105*, 221 (1943)

10. Bajaj, Y. P. S.: *In*: Applied and Fundamental Aspects of Plant Cell, Tissue and Organ Culture. Reinert, J., Bajaj, Y. P. S. (Eds.), p. 467. Springer Verlag 1977
11. Barg, R., Umiel, N.: Z. Pflanzenphysiol. *83*, 437 (1977)
12. Basu, R., Roy, B. N., Bose, T. K.: Plant and Cell Physiol. *11*, 681 (1970)
13. Bernstein, L.: Ann. Rev. Phytopathology *13*, 295 (1975)
14. Binding, H., Binding, K., Straub, J.: Naturwissenschaften *57*, 138 (1970)
15. Bitters, W. P., Murashige, T.: Calif. Citrograph *52*, 226, 270, 304 (1970)
16. Bitters, W. P., Murashige, T., Rangan, T. S., Nauer, E.: Calif. Citrus Nurserymen's Soc. *9*, 27 (1970)
17. Blaydes, D. F.: Physiologia Plant. *19*, 748 (1966)
18. Bottino, P. J.: Radiation Bot. *15*, 1 (1975)
19. Broertjes, C., Haccius, B., Weidlich, S.: Euphytica *17*, 321 (1968)
20. Brunet, G., Ibrahim, R. K.: Z. Pflanzenphysiol. *69*, 152 (1973)
21. Button, J., Bornman, C. H.: J. S. Afr. Bot., 37, 127 (1971)
22. Button, J., Botha, C. E. J.: J. Expt. Bot. *26*, 723 (1975)
23. Button J., Kochba, J.: *In*: Applied and Fundamental Aspects of Plant Cell, Tissue and Organ Culture. Reinert, J., Bajaj, Y. P. S., (Eds.), p. 70. Springer Verlag, Heidelberg, Berlin 1977
24. Button, J., Kochba, J., Bornman, C. H.: J. Exptl. Botany *25*, 446 (1974)
25. Button, J., Kochba, J., Bornman, C. H., Gilliland, H. G., Evers, P.: Proc. Electron Microscopy Soc. of S. Africa *3*, 5 (1973)
26. Button, J., Vardi, A., Spiegel-Roy, P.: Theor. Appl. Gen. *47*: 119 (1975)
27. Cameron, J. W., Frost, H. B.: *In*: Citrus Industry. Reuther, W., Webber, H. J., Batchelor, L. D. (Eds.) Vol. II, p. 325. Div. Agric. Sci. University of California 1968
28. Cameron, J. W., Soost, R. K.: *In*: Outlines of perennial crop breeding in the tropics. Ferwerda, F. P., Wit, F. (Eds.), p. 129. Wageningen, Veenman and Zonen 1969
29. Carlson, P. S.: Science *180*, 1366 (1973)
30. Chaleff, R. S., Carlson, P. S.: Annu. Rev. Genet. *8*, 267 (1974)
31. Chapman, H. D.: *In*: The Citrus Industry. Reuther, W., Batchelor, L. D., Webber, H. J. (Eds.), Vol. II, p. 127. Div. Agric. Sciences, Univ. of California 1968
32. Chourey, P. S., Smith, H. H., Combatti, N. C.: Amer. J. Bot. *60*, 853 (1973)
33. Colclasure, G. C., Yopp, J. H.: Physiol. Plantarum *37*, 298 (1976)
34. Cooper, W. C., Gorton, B. S., Edwards, C.: Rio Grande Val. Hort. Proc. *5*, 46 (1951)
35. Davies, P.: Plant Physiol. *57*, 192 (1976)
36. De Lange, J. H., Vincent, A. P.: *In*: 1977 Proc. Int. Soc. Citriculture. Grierson, W. (Ed.), Vol. II, p. 569. Lake Alfred, Florida 1979
37. Dix, P. J., Street, H. E.: Plant Sci. Lett. *5*, 231 (1975)
38. Epstein, E., Kochba, J., Neumann, H.: Z. Pflanzenphysiol. *85*: 263 (1977)
39. Erner, Y., Reuveni, O., Goldschmidt, E. E.: Plant Physiol. 56, 279 (1975)
40. Esan, E. B.: *In*: A detailed study of adventive embryogenesis in the Rutaceae. Ph. D. dissertation, Univ. of California, Riverside 1973
41. Esen, A., Scora, R. W., Soost, R. K.: J. Am. Soc. Hort. Sci. *100*, 558 (1975)
42. Esen, A., Soost, R. K.: J. Hered. *67*, 199 (1976)
43. Frost, H. B., Soost, R. K.: *In*: The Citrus Industry. Reuther, W., Batchelor, L. D., Webber, H. J., (Eds.), Vol. II, p. 290. Univ. of Calif., Division of Agr. Sciences, Berkeley 1965
44. Fujimura, T., Komamine, A.: Plant Sci. Lett. *5*, 359 (1975)
45. Furusato, K., Ohta, Y., Ishibashi, K.: Biol. Res. *8*, 40 (1957)
46. Galston, A. W., Davies, P. J.: Science *163*, 1288 (1969)
47. Giles, K. L.: *In*: Applied and Fundamental Aspects of Plant Cell, Tissue and Organ Culture. Reinert, J., Bajaj, Y. P. S., (Eds.), p. 536. Springer Verlag 1977
48. Goldner, R., Umiel, N., Chen, Y.: Z. Pflanzenphysiol. *85*, 307 (1977)
49. Göring, H., Reckin, E.: Flora, Abt. A. *159*, 82 (1968)
50. Gressel, J., Zilkah, S., Ezra, G.: *In*: Proc. Fourth Int. Plant Tissue Culture Congress, Calgary, Canada (in press) 1978
51. Grinblat, U.: J. Amer. Soc. Hort. Sci. *97*, 599 (1972)
52. Halperin, W., Wetherell, D. F.: Amer. J. Bot. *51*, 274 (1964)
53. Harari, D., Umiel, N.: Israel J. Bot. *24*, 55 (1975)

54. Hausenberg, Y., Rozen, Y., Boaz, M.: An Israeli Ministry of Agriculture Publication (in Hebrew) Sept. 1968
55. Hearn, C. J.: Proc. Fla. St. Hort. Soc. *86*, 84 (1974)
56. Hess, D.: *In*: Applied and Fundamental Aspects of Plant Cell, Tissue and Organ Culture. Reinert, J., Bajaj, Y. P. S., (Eds.), p. 506. Springer 1977
57. Hildebrandt, A. C., Riker, A. J.: Am. J. Bot. *36*, 74 (1949)
58. Hoffmann, F., Kull, U., Jeremias, K.: Z. Pflanzenphysiol. *64*, 223 (1971)
59. Iglesias, L., Lima, H., Simon, J. P.: J. Hered. *65*, 81 (1974)
60. Ikeda, F.: *In*: Improvement of Vegetatively Propagated Plants and Tree Crops through Induced Mutations. IAEA Bull., p. 95. Wageningen 1976
61. Isikawa, H.: Bot. Mag. Tokyo *87*, 73 (1974)
62. Iwamasa, M., Ueno, I., Nishiura, M.: Bull. Hort. Res. Sta., Japan, Ser. B, No. 7 (1967)
63. Jeffs, R. A., Northcote, D. H.: J. Cell Sci. *2*, 77 (1967)
64. Johri, B. M., Ahuja, M. R.: Curr. Sci. *25*, 162 (1956)
65. Juarez, J., Navarro, L., Guardiola, J. L.: Fruits *31*, 751 (1976)
66. Kim, C. M., Kim, J. K., Kim, H. W., Moon, J. D.: Korean J. Breed. *4*, 132 (1972)
67. Kochba, J., Button, J.: Z. Pflanzenphysiol. *73*, 415 (1974)
68. Kochba, J., Button, J., Spiegel-Roy, P., Bornman, C. H., Kochba, M.: Ann. Botany *38*, 111 (1974)
69. Kochba, J., Lavee, S., Spiegel-Roy, P.: Plant and Cell Physiol. *18*, 463 (1977)
70. Kochba, J., Spiegel-Roy, P.: Z. Pflanzenzucht *69*, 156 (1973)
71. Kochba, J., Spiegel-Roy, P.: The Plant Propagator *22*, 11 (1976)
72. Kochba, J., Spiegel-Roy, P.: *In*: Proc. Res. Meeting, IAEA Bull. 194, p. 83. Wageningen, Holland 1976
73. Kochba, J., Spiegel-Roy, P.: Hort. Sci. *12*, 110 (1977)
74. Kochba, J., Spiegel-Roy, P.: Environment. and Exptl. Botany *17*, 151 (1977)
75. Kochba, J., Spiegel-Roy, P.: Pflanzenphysiol. *81*, 283 (1977)
76. Kochba, J., Spiegel-Roy, P., Neumann, H., Saad, S.: Z. Pflanzenphysiol. *89*, 427 (1978)
77. Kochba, J., Spiegel-Roy, P., Saad, S.: Proc. Res. Coord. Meeting, Skiernewietze, Poland. IAEA Bull. (in press) 1978
78. Kochba, J., Spiegel-Roy, P., Saad, S., Neumann, H.: Acta Horticulturae, *78*, 185 (1977)
79. Kochba, J., Spiegel-Roy, P., Saad, S., Neumann, H.: *In*: Proc. Int. Symp. Plant Cell Culture. Alfermann, A. W., Reinhard, E., (Eds.), p. 223. GSF, München 1978
80. Kochba, J., Spiegel-Roy, P., Saad, S., Neumann, H.: Naturwissenschaften *65*, 261 (1978)
81. Kochba, J., Spiegel-Roy, P., Safran, H.: Planta *106*, 237 (1972)
82. Krul, W. R., Colclasure, G. C.: Physiol. Plantarum *41*, 135 (1977)
83. Lavee, S., Galston, A. W.: Am. J. Bot. *55*, 890 (1968)
84. Le Page-Degivry, M. Th.: Z. Pflanzenphysiol. *70*, 406 (1973)
85. Lescure, A. M.: Plant Sci. Lett. *1*, 375 (1973)
86. Lin, B. C., Kado, C. I.: Bot. Bull. Acad. Sinica *18*, 71 (1977)
87. Maheshwari, P., Rangaswamy, N. S.: Indian J. Hort. *15*, 275 (1958)
88. Malca, I., Endo, R. M., Long, M. R.: Phytopathology *57*, 272 (1967)
89. Maliga, P.: *In*: Cell Genetics in Higher Plants. Dudits, D., Farkas, G. L., Maliga, P., (Eds.), p. 59. Hung. Acad. Sci. 1976
90. Maliga, P., Marton, L., Breznovits, A. S.: Plant Sci. Lett. *1*, 119 (1973)
91. Maretzki, A., Thom, M.: Plant Physiol. *61*, 544 (1978)
92. Masuda, K., Koda, Y., Okazawa, Y.: Physiol. Plantarum *41*, 135 (1977)
93. Melchers, G.: Z. Pflanzenzucht. *67*, 19 (1972)
94. Melchers, G., Bergmann, L.: Ber. dtsch. Bot. Ges. *71*, 459 (1958)
95. Melchers, G., Labib, G.: Ber. dtsch. Bot. Ges. *83*, 129 (1970)
96. Milborrow, B. V.: Planta (Berl.) *70*, 155 (1966)
97. Milborrow, B. V.: Annu. Rev. Plant Physiol. *25*, 259 (1974)
98. Mitra, G. C., Chaturvedi, H. C.: Bull. Torrey Bot. Cl. *99*, 184 (1972)
99. Murashige, T.: Ann. Rev. Plant Physiol. *25*, 135 (1974)
100. Murashige, T., Bitters, W. P., Rangan, T. S., Nauer, E. M., Roistacher, C. N., Holliday, P. B.: Hort Sci. *7*, 118 (1972)
101. Murashige, T., Skoog, F.: Physiol. Plant. *15*, 473 (1962)

102. Murashige, T., Tucker, D. P. H.: *In*: Proc. 1st. Int. Citrus Symp. Vol. *3*, p. 1155. Univ. of Calif., Riverside 1969
103. Nabors, M. W., Daniels, A., Nadolny, L., Brown, C.: Plant Sci. Lett. *4*, 155 (1975)
104. Nagata, T., Takebe, I.: Planta *99*, 12 (1971)
105. Navarro, L.: *In*: Proc. Int. Soc. Citriculture. Grierson, W. (Ed.), Vol. *1*, p. 136. 1977
106. Navarro, L., Roistacher, C. N., Murashige, T.: J. Amer. Soc. Hort. Sci. *100*, 471 (1975)
107. Newcomb, W., Wetherell, D. F.: Bot. Gaz. *131*, 242 (1970)
108. Nickell, L. G., Maretzki, A.: Plant and Cell Physiol. *11*, 183 (1970)
109. Nishiura, M., Iwamasa, M.: Mull. Hort. Res. Sta., Japan, Ser. B, No. **10** (1970)
110. Nishiura, M., Iwamasa, M., Ueno, I.: Bull. No. *1*, Hort. Res. Sta., Okitsu, Japan, Ser. B., 1974
111. Nordby, H. E., Nagy, S.: Proc. Fla. St. Hort. Soc. *87*, 70 (1974)
112. Ohata, Y., Furusato, K.: Seiken Ziho, *8*, 49 (1957)
113. Ordin, L., Bonner, J.: Plant Physiol. *32*, 212 (1957)
114. Ozsan, M., Cameron, J. W.: Proc. Amer. Soc. Hort. Sci. *82*, 210 (1963)
115. Parlevliet, J. E., Cameron, J. W.: Proc. Amer. Soc. Hort. Sci. *74*, 252 (1959)
116. Potrykus, I.: Z. Pflanzenphysiol. *70*, 364 (1973)
117. Potrykus, I., Hoffmann, F.: Z. Pflanzenphysiol. *69*, 287 (1973)
118. Power, J. B., Cocking, E. C.: *In*: Applied and Fundamental Aspects of Plant Cell, Tissue and Organ Culture. Reinert, J., Bajaj, Y. P. S., (Eds.), p. 497. Springer Verlag 1977
119. Raa, J.: Physiologia Plant. *24*, 498 (1971)
120. Raa, J.: Physiologia Plant. *25*, 130 (1971)
121. Rangan, T. S., Murashige, T., Bitters, W. P.: Hort Sci. *3*, 226 (1968)
122. Rangan, T. S., Murashige, T., Bitters, W. P.: *In*: Proc. 1st. Int. Citrus Symp. Vol. I, p. 225. Univ. of California, Riverside 1969
123. Rangaswamy, N. S.: Nature *183*, 735 (1959)
124. Rangaswamy, N. S.: Phytomorphology *11*, 109 (1961)
125. Raveh, D., Galun, E.: Z. Pflanzenphysiol. *76*, 76 (1975)
126. Raveh, D., Huberman, E., Galun, E.: In Vitro *9*, 216 (1973)
127. Reinert, J., Tazawa, M.: Planta (Berl.) *87*, 239 (1969)
128. Roberts, R. M., Heishman, A., Wicklin, C.: Plant Physiol. *48*, 36 (1971)
129. Rush, D. W., Epstein, E.: Plant Physiology *57*: 162 (1976)
130. Sabharwal, P. S.: *In*: Plant tissue and organ culture. Maheshwari, P., Rangaswamy, N.S., (Eds.), p. 102. Univ. of Delhi, India 1963
131. Salomon, E., Mendel, K.: Am. Soc. Hort. Sci. *86*: 213 (1965)
132. Scandalios, J. G.: Ann. Rev. Plant Physiol. *25*, 225 (1974)
133. Siegel, B. Z., Galston, A. W.: Plant Physiol. *42*, 221 (1967)
134. Singh, U. R.: *In*: Plant Tissue and Organ Culture. a Symposium. Maheswari, P., Rangaswamy, N. S. (Eds.), p. 275. Delhi 1963
135. Spiegel-Roy, P.: On the chimeral nature of Shamouti orange. Euphytica *28*, 361 (1979)
136. Spiegel-Roy, P., Kochba, J.: Radiation Bot. *13*, 97 (1973)
137. Spiegel-Roy, P., Kochba, J.: *In*: Induced Mutations in Vegetatively Propagated Crops. Proc. of a Panel, IAEA Publ. No. 339, p. 91. Vienna 1973
138. Spiegel-Roy, P., Kochba, J.: Proc. Res. Meeting, Tokai, Japan. IAEA Bull. *173*, 117 (1975)
139. Spiegel-Roy, P., Kochba, J.: *In*: Res. Meeting on the Use of Ionizing Radiations in Agriculture. p. 441. Euratom-ITAL, Wageningen, Holland 1976
140. Spiegel-Roy, P., Kochba, J.: *In*: Plant Tissue Culture and its Biotechnological Application. Proc. 1st. Int. Congr. Medicinal Plants. Barz, W., Reinhard, E., Zenk, M. H. (Eds.), Section *B*, p. 404. Springer Verlag 1977
141. Spiegel-Roy, P., Teich, A. H.: Euphytica *31*, 534 (1972)
142. Spiegel-Roy, P., Vardi, A., Shani, A.: *In*: 1977 Proc. Int. Soc. Citriculture. Grierson, W. (Ed.), Vol. II, p. 619, Lake Alfred, Florida 1979
143. Stenlid, G.: Physiol. Plant. *12*, 218 (1959)
144. Steward, F. C., Ammirato, P. W.: Ann. Bot. *34*, 761 (1970)
145. Steward, F. C., Mapes, M. O., Ammirato, P. N.: *In*: Plant Physiology: A Treatise. Steward, F. C. (Ed.), p. 329. London, N.Y. Academic Press 1969
146. Sung, R. Z.: Genetics *84*, 51 (1976)

147. Swingle, W. T.: *In*: The Citrus Industry. Reuther, W., Webber, H. J., Batchelor, L. D. (Eds.), Vol. I, p. 190. Div. Agric. Sci., University of California 1967
148. Tal, M.: Austr. J. agric. Res. *22*, 631 (1971)
149. Tatum, J. H., Berry, R. E., Hearn, C. J.: Proc. Fla. St. Hort. Soc. *87*, 75 (1974)
150. Tatum, J. H., Hearn, C. J., Berry, R. E.: J. Amer. Soc. Hort. Sci. 103, 492 (1978)
151. Thomas, E., Street, H. E.: Ann. Bot. *34*, 675 (1970)
152. Thorpe, T. A., Maier, V. P., Hasegawa, S.: Phytochemistry *10*, 711 (1971)
153. Tisserat, B., Murashige, T.: In Vitro *13*, 785 (1977)
154. Tisserat, B., Murashige, T.: In Vitro *13*, 799 (1977)
155. Toxopeus, H. J.: Züchter *8*, 1 (1936)
156. Ueno, I., Nishiura, M.: Bull. Hort. Res. Sta., Japan, Ser. B, No. *9* (1969)
157. Van Hoof, P., Gaspar, T. H.: Scientia Hort. *4*, 27 (1976)
158. Vardi, A.: *In*: Proc. Intern. Symposium on Plant Cell Culture. Alfermann, A. W., Reinhard, E. (Eds.), p. 234. BPT-Report 1/78 (1978)
159. Vardi, A., Spiegel-Roy, P.: *In*: 1978 Proc. Int. Soc. Citriculture. Grierson, W. (Ed.). Lake Alfred, Florida (in press) 1979
160. Vardi, A., Spiegel-Roy, P., Galun, E.: Plant Sci. Lett. *4*, 231 (1975)
161. Vardi, A., Raveh, D.: Z. Pflanzenphysiol. *78*, 350 (1976)
162. Wareing, P. F., Good, J., Potter, H., Pearson, A.: *In*: Plant Growth Regulators. Monograph 31, p. 191. Soc. Chem. Ind. London 1968
163. White, P. R.: *In*: The Cultivation of Animal and Plant Cells. New York, Ronald Press 1963
164. Widholm, J. M.: *In*: Plant Tissue Culture and its Biotechnological Applications. Barz, W., Reinhard, E., Zenk, M. H. (Eds.), p. 112. Springer Verlag 1976
165. Wolter, K. E., Gordon, J. C.: Physiol. Plant. *33*, 219 (1975)
166. Zenk, M. H.: *In*: Haploids in Higher Plants. Kasha, K. (Ed.), p. 339. Univ. of Guelph, Ontario, Canada 1974
167. Zilkah, S., Bocion, P. F., Gressel, J.: Plant and Cell Physiol. *18*, 657 (1977)
168. Zilkah, S., Gressel, J.: Plant and Cell Physiol. *18*, 815 (1977)

Biotransformation by Plant Cell Cultures

E. Reinhard, A. W. Alfermann
Lehrstuhl für Pharmazeutische Biologie der Universität Tübingen
D-7400 Tübingen, Federal Republic of Germany

Dedicated to Prof. Dr. W. Simonis (Würzburg) on the occasion of his 70th birthday

The biotransformation patterns of various phenols, aniline derivatives, benzoic and cinnamic acids, coumarins, tropane, furoquinoline, isoquinoline and ergot alkaloids, mono-, di- and triterpenoids, pregnane and androstane derivatives, cholesterol and cardenolides by plant cell cultures are reviewed. Using biotransformation of cardiac glycosides as an example it is shown that the pattern of biotransformation products obtained depends on the special cell strain used. It is demonstrated that it is possible to isolate cell strains with a high biotransformation capacity of a special desired type, namely hydroxylation of β-methyldigitoxin to β-methyldigoxin. Practical application of biotransformation by plant cell cultures is discussed.

1 Introduction

Higher plants are still important sources of medicinal or chemical compounds. In recent years, however, supplies of these plants have become difficult to maintain. Therefore, increasing interest has focused on the possibility to produce secondary products by plant cell cultures. Although a technical application of this method has not yet been achieved, very promising progress has been made during the last decade[1-5].

It has been shown that plant cell cultures are able to perform special biotransformation reactions on organic compounds added to the medium. These compounds may or may not be common to the plant species from which the cell culture has been derived. Therefore, it becomes possible to transform a substance of a lower to a higher scientific or commercial value and eventually up to now not available in sufficient amounts. Furthermore, substances of unknown structure and new chemical and pharmacological properties can be synthesized.

This paper summarizes the results obtained up to now and discusses those chemical biotransformations in plant cell cultures (suspensions) in which compounds of increased complexity are formed. Biotransformations of a degradative type are dealt with elsewhere[6,7].

2 Aromatic Compounds

2.1 Phenols (Fig. 1)

Biotransformation of ortho-, meta- and paradihydroxybenzene (catechol, resorcinol, hydroquinone and vanillin) to the corresponding β-glucosides by cell cultures of *Agrostemma githago*, *Datura ferox* and *Digitalis purpurea* was first reported by Pilgrim in 1970[8]. Additionally salicyl alcohol and salicyl aldehyde are transformed by these cells to o-hydroxybenzyl-β-D-glucoside (isosalicin), whereas salicin was not detected. Helicin was found only in trace amounts after feeding of salicyl aldehyde. These results have been confirmed by Tabata *et al.*[9] using cell cultures of *Datura innoxia*, except for the formation of helicin. Arbutin formation after addition of hydroquinone has been investigated in more detail by these authors. Glucosylation proceeded quickest, when the substrate was added during the linear growth phase and 10^{-3} M of hydroquinone was glycosylated during 10 h of incubation. Starting with incubation on day 7 (10^{-3} M) and adding new substrate (5×10^{-4} M) every day for five further days, 70% of the added hydroquinone was glucosylated to arbutine within 6 days, although the growth of the cells was reduced remarkably. Apparent degradation of the product could be avoided, if the biotransformation was achieved during the stationary phase. The biotransformation rate, however, was significantly lower. First studies with crude enzyme extracts showed that UDPG served as a high energy donor of glucose.

2.2 Acetanilide, Aniline, Anisole, Benzoic Acid (Fig. 2)

Besides biotransformation of coumarin that of aniline, anisole, acetanilide and benzoic acid by cell cultures of *Catharanthus roseus*, *Conium maculatum* and

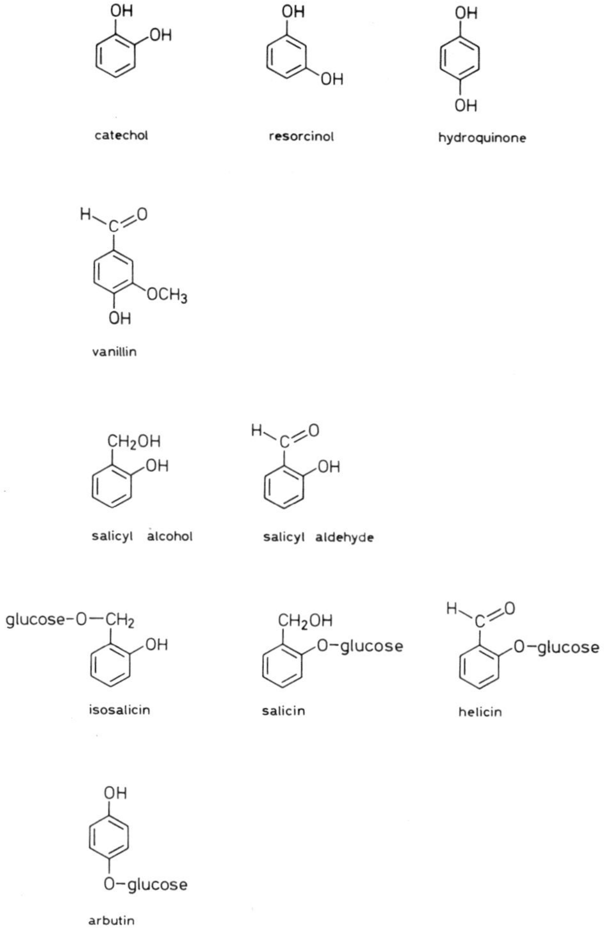

catechol resorcinol hydroquinone

vanillin

salicyl alcohol salicyl aldehyde

isosalicin salicin helicin

arbutin

Fig. 1

Apocynum cannabinum has been investigated by Carew and Bainbridge[11]. Aniline was hydroxylated to 3-hydroxyaniline or acetylated to acetanilide, respectively, by *Catharanthus* cultures depending on the growth phase of the cells, whereas with *Conium* and *Apocynum* only acetanilide was found. Benzoic acid was hydroxylated to 4-hydroxybenzoic acid by all three cell cultures; anisole was demethylated to phenol by *Catharanthus*, whereas acetanilide did not undergo further biotransformation.

2.3 Cinnamic Acid Derivatives (Fig. 3)

Steck and Constable[10] incubated ^{14}C-labelled ortho-, meta- and parafluoro- as well as -methylcinnamic acids with cell cultures of *Ammi visnaga* and *Nicotiana tabacum* and found that fluorine atoms were fairly readily removed from the aromatic ring in course of metabolism, whereas the methyl groups were not. Tobacco gave higher biotransformation rates than *Ammi* and the following metabolites were found: ^{14}C

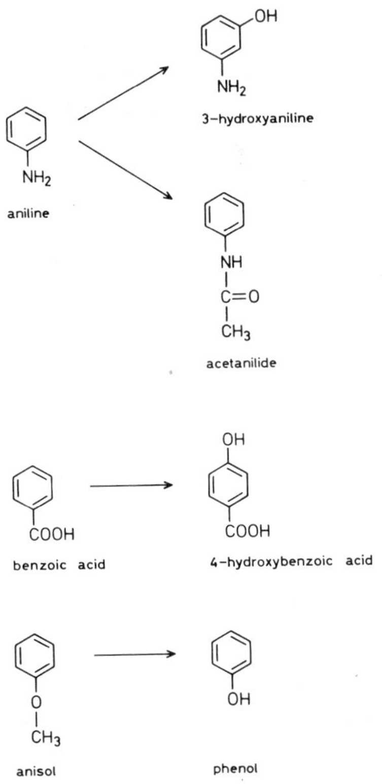

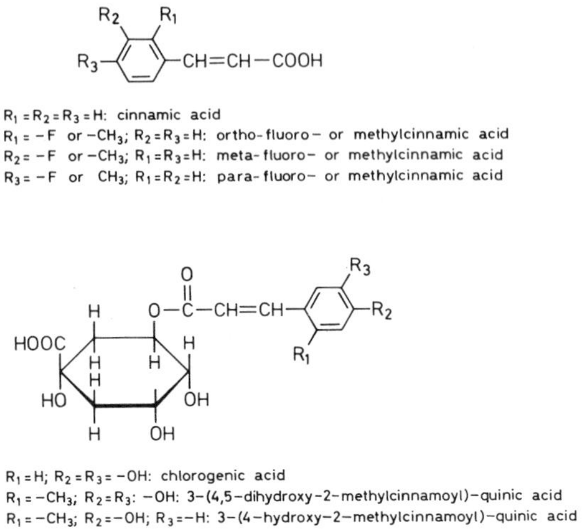

aniline → 3-hydroxyaniline; aniline → acetanilide

benzoic acid → 4-hydroxybenzoic acid

anisol → phenol

◀ **Fig. 2** Tranformation of aniline, benzoic acid and anisole by cell cultures of *Catharanthus roseus*, *Conium maculatum* and *Apocynum cannabinum*

$R_1 = R_2 = R_3 = H$: cinnamic acid
$R_1 = -F$ or $-CH_3$; $R_2 = R_3 = H$: ortho-fluoro- or methylcinnamic acid
$R_2 = -F$ or $-CH_3$; $R_1 = R_3 = H$: meta-fluoro- or methylcinnamic acid
$R_3 = -F$ or CH_3; $R_1 = R_2 = H$: para-fluoro- or methylcinnamic acid

$R_1 = H$; $R_2 = R_3 = -OH$: chlorogenic acid
$R_1 = -CH_3$; $R_2 = R_3$: $-OH$: 3-(4,5-dihydroxy-2-methylcinnamoyl)-quinic acid
$R_1 = -CH_3$; $R_2 = -OH$; $R_3 = -H$: 3-(4-hydroxy-2-methylcinnamoyl)-quinic acid

Fig. 3 Transformation products of cinnamic acid derivatives by cell cultures of *Nicotiana tabacum* and *Ammi visnaga*

was found in chlorogenic acid and scopoletin, indicating that the fluoro substrates underwent dehalogenation; labelled methylcinnamic acids were hydroxylated by tobacco. The biotransformation of ortho-methylcinnamic acid represents an interesting example of unnatural product formation: this substrate was transformed to the quinic ester of 4,5-dihydroxy-2-methyl- and 4-hydroxy-2-methylcinnamic acid, respectively.

3 Coumarins (Fig. 4)

Coumarin itself is hydroxylated by cell cultures of *Cantharanthus roseus* to 7-hydroxycoumarin[11] which serves in *Ruta* cell cultures as a precursor of marmesin, herniarin and psoralen[10]. Normally, the substrate specificity of the enzymes involved in secondary product formation is very high, and unnatural substrates are not accepted. However, this is not always the case, and especially peroxidases, alcohol dehydrogenases, methyltransferases, or glucosidases are known to have no very high substrate specificity. Therefore, Steck and Constable[10] tested the possibility of biotransformation of analogs of 7-hydroxycoumarin, like 4-methyl- and 8-methyl-7-hydroxycoumarin, by cells of *Ruta graveolens*. And, indeed, they found as biotransformation products

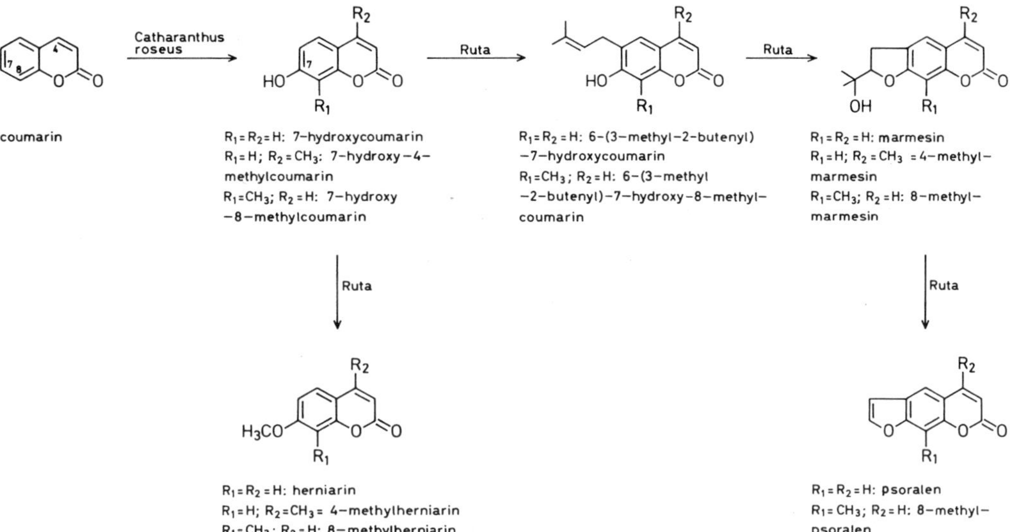

Fig. 4 Transformation of coumarin and its derivatives by cell cultures of *Catharanthus roseus* and *Ruta graveolens*

unnatural compounds like the 4- and 8-methyl analogs of herniarin and marmesin along with 6-(3-methyl-2-butenyl)-7-hydroxy-8-methyl-coumarin and, probably, 8-methylpsoralen.

4 Alkaloids

4.1 Tropane Alkaloids (Fig. 5)

Cell cultures derived from plants containing tropane alkaloids often cease to synthesize these alkaloids. Therefore, biotransformation of alkaloid precursors has been tested. On a 2,4-D containing medium, cell cultures of *Datura innoxia*[12, 13], *Datura metel*[12], *Datura ferox*[14], *Datura stram. ssp. stramonium*[12] and *Datura stram. ssp. tatula*[13] esterify exogenously supplied tropine not with tropic acid but with endogenous acetic acid to acetyltropine. On a medium containing 2 mg l^{-1} NAA

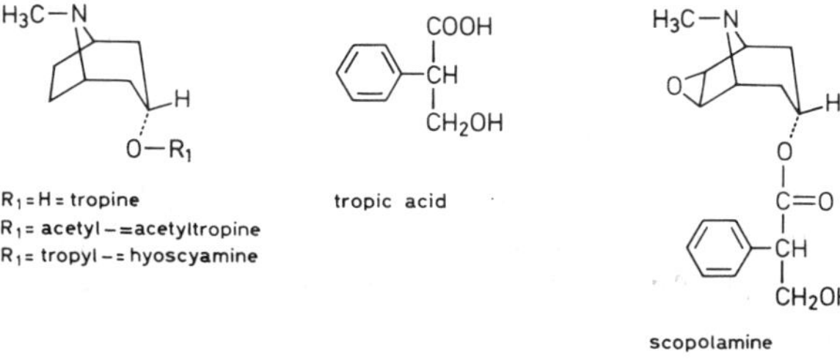

Fig. 5 Transformation products of tropane alkaloid precursors by cell cultures of *Datura sp.*

instead of 2,4-D Romeike[12] observed formation of hyoscyamine (atropine). However, simultaneous formation of scopolamine by undifferentiated cell cultures of *Datura innoxia* was never found. Both reactions have been reported by Stohs[15] for cell cultures of *Datura stramonium* and by Indra and Staba[16] for cell free extracts of that plant.

4.2 Furoquinoline Alkaloids (Fig. 6)

Precursor biotransformation of furoquinoline alkaloids has been done by Steck *et al.*[17] and Steck and Constable[10] using cell cultures of rue. They could show that 4-hydroxy-2-quinolone was specifically used as a precursor of dictamnine which is only present in these cell cultures after feeding this appropriate precursor. When the quinolone concentration of the medium decreases, dictamnine is transformed further to 8-methoxydictamnine.

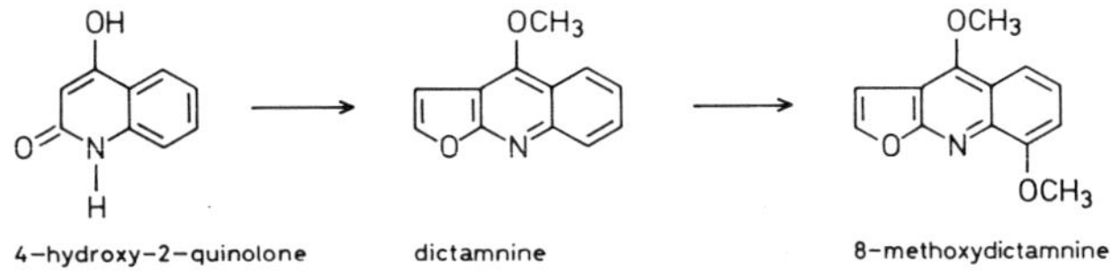

OH OCH3 OCH3

4-hydroxy-2-quinolone dictamnine 8-methoxydictamnine

Fig. 6 Transformation of 4-hydroxy-2-quinolone by cell cultures of *Ruta graveolens*

4.3 Isoquinoline Alkaloids

Cell cultures of *Papaver sp.* stop to synthesize the important alkaloids of this plant species as i.e. morphine, codeine or thebaine, but are still able to perform special biotransformation reactions on alkaloids added into the medium[18,19] and also isolated latex from the capsules has been shown to contain biological activity[20]. From (RS)-reticuline, only (+)-(S)-reticuline was stereospecifically transformed to (+)-(S)-scoulerine (14.7% of 100 mg l^{-1}) and to (S)-cheilanthifoline (0.5%), whereas (—)-(R)-reticuline (9.6%) remained unmetabolized[19] (Fig. 7). No morphinane alkaloids could be detected. Therefore, the loss of morphinane alkaloid formation in *Papaver* cells can be attributed to a loss of phenol oxidation enzyme catalyzing the reaction of (—)-(R)-reticuline to salutaridine[19], whereas the berberine bridge enzyme also recently described for callus cultures of *Macleaya microcarpa*[21] converting (+)-(S)-reticuline to scoulerine is present. Thebaine and morphine were not transformed in these investigations[19]. However (—)-codeinone was reduced to (—)-codeine. Biotransformation of thebaine to codeine has been reported in older investigations by Grützmann and Schröter[18] for cell cultures of *Papaver somniferum* and even *Nicotiana alata* (Fig. 7).

4.4 Indole Alkaloids

Biotransformation of vindoline by a *Catharanthus roseus* crown gall cell culture had been investigated already in 1964[22] and, besides unknown oxidation products, desacetylvindoline could be identified. Additionally, Carew and Krueger[23] found

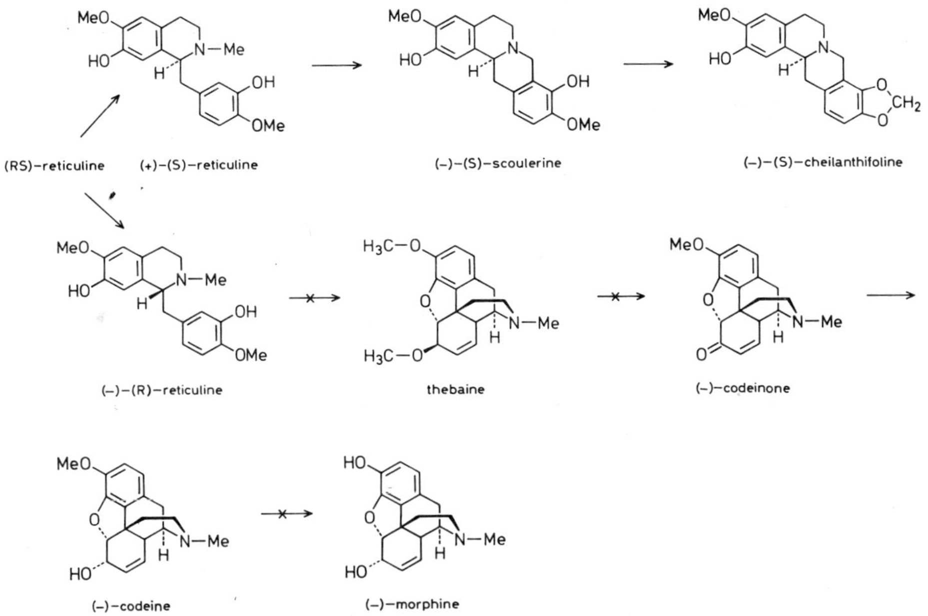

(RS)-reticuline (+)-(S)-reticuline

(–)-(S)-scoulerine

(–)-(S)-cheilanthifoline

(–)-(R)-reticuline

thebaine

(–)-codeinone

(–)-codeine

(–)-morphine

Fig. 7 Transformation of (RS)-reticuline by cell cultures of *Papaver somniferum*

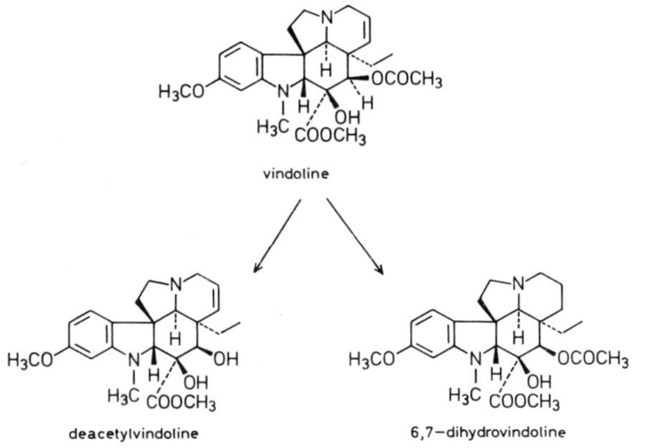

vindoline

deacetylvindoline

6,7-dihydrovindoline

Fig. 8 Transformation of vindoline by cell cultures of *Catharanthus roseus*

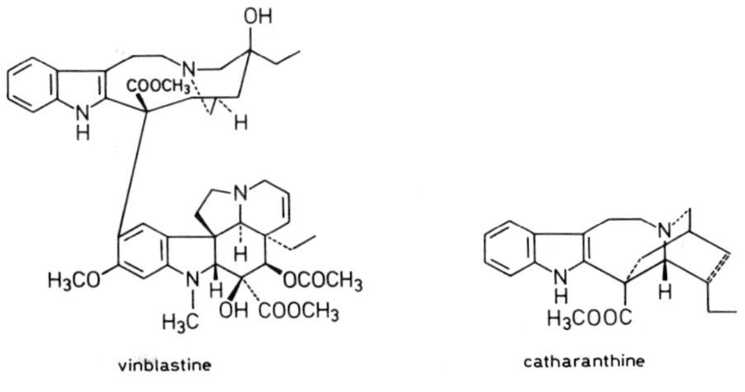

vinblastine

catharanthine

Fig. 9

dihydrovindoline as conversion product (Fig. 8). No metabolite of catharanthine was detected, and after feeding of both vindoline and catharanthine no dimeric alkaloids were found. Vinblastine was metabolized to three unidentified compounds, two of which were dimeric alkaloids according to MS spectroscopy (Fig. 9).

Feeding of tryptophane to cell cultures of *Phaseolus vulgaris* leads to the formation of harman and norharman (Fig. 10), otherwise unknown for this species[24, 25].

harman norharman

Fig. 10 Transformation products of tryptophan by cell cultures of *Phaseolus vulgaris*

4.5 Ergot Alkaloids

Agroclavine and elymoclavine (Fig. 11) were incubated with suspension cultures of *Anethum graveolens*, *Allium cepa*, *Withania somnifera* and *Capsicum annuum*. All of these tissues were able to hydroxylate both compounds at C-8, none were capable of transforming agroclavine to elymoclavine. However, depending on the cell culture used, appropriate degradation of substrates and products was observed[26].

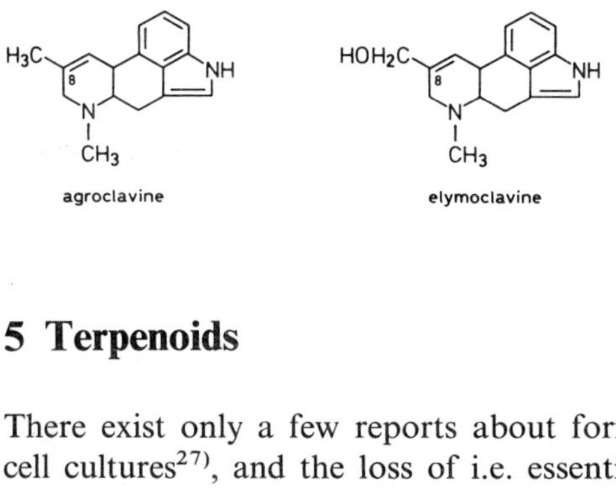

agroclavine elymoclavine

Fig. 11 Substrates used in biotransformation with cell cultures of *Anethum graveolens*, *Allium cepa*, *Withania somnifera* and *Capsicum annuum*

5 Terpenoids

There exist only a few reports about formation of terpenoids by undifferentiated cell cultures[27], and the loss of i.e. essential oil formation in many *Lamiaceae* cell cultures is thought to be due to the fact that secondary product formation is coupled with morphogenetic differentiation as in the example mentioned with differentiation of oil cells and glandules. Therefore, the ability of certain (undifferentiated) cell cultures to metabolize various terpenoids is of special interest.

5.1 Monoterpenoids

No cannabinoids are synthesized even after feeding of geraniol, nerol, olivetol or ethyl olivetol, but some of these precursors are oxidized by cell cultures of *Cannabis sativa* as well as by cell-free extracts thereof[28, 29, 30]. Geraniol and nerol are transformed to citral a + citral b (35.2%) and nerol (14.8%) or citral a + citral b (46.8%) and geraniol (8.2%), respectively (Fig. 12). The conversion of geraniol

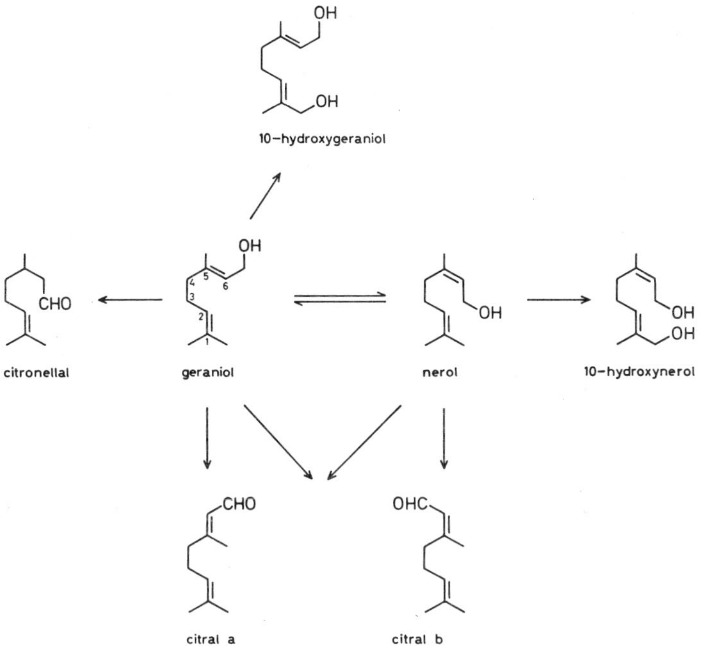

Fig. 12 Transformation of geraniol and nerol by cell cultures of *Cannabis sativa*, *Rosa sp.* and *Catharanthus roseus*

to nerol was observed only occasionally and could not be found with cell free extracts. The enzyme involved is obviously an alcohol oxidase; besides an oxidation of cis- and trans-verbenol to verbenone (Fig. 13), this enzyme can also oxidize alcohols of different structures like cinnamyl alcohol to cinnamaldehyde, isophorol to isophoron, or 3-phenyl-1-propanol to the appropriate propanal. However, no biotransformation of citronellol, borneol, menthol, α- and β-ionol or linalool was achieved with this system. Formation of citral a and b and nerol, but additionally also of citronellal has been found after incubation of geraniol and geranyl-pyrophosphate with *Rosa* cultures[31] (Fig. 12). Furthermore, a microsomal mixed function oxidase from *Catharanthus roseus* seedlings has been described to hydro-

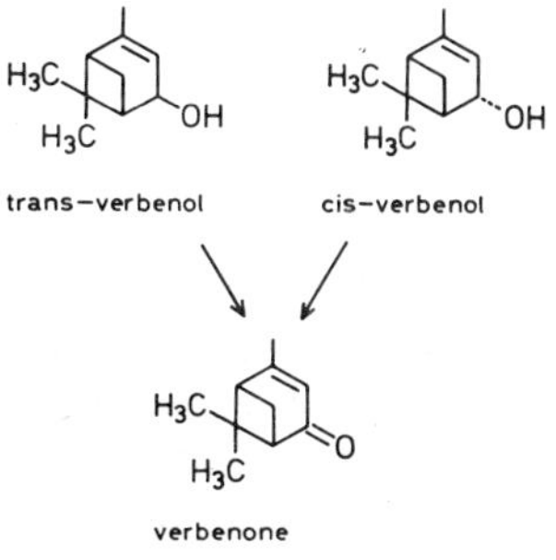

Fig. 13 Transformation of cis- and trans-verbenol by cell-free extracts of *Cannabis sativa* cell cultures

xylate geraniol and nerol to the corresponding 10-hydroxy derivatives[32] (Fig. 12). In contrast to *Cannabis* and *Rosa*, cell cultures of tobacco are able to transform linalool, and its derivatives, dihydrolinalool, linalyl acetate and dihydrolinalyl acetate at C-8 to the 8-hydroxyderivatives (Fig. 14). Furthermore, linalyl acetate as well as dihydrolinalyl acetate are hydrolysed and hydroxylated, so that 8-hydroxylinalool and 8-hydroxydihydrolinalool occur[33]. Cell cultures of *Mentha* are able to reduce (+)-pulegon to (+)-isomenthone (Fig. 15), and it could be shown that the capacity of the cell strain to perform or not to perform this reaction depends on the genetic constitution of the plant from which the cell culture has been derived[34].

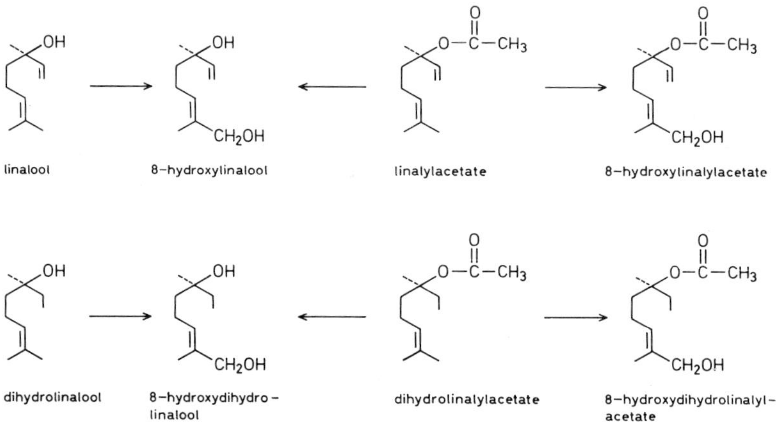

linalool → 8-hydroxylinalool ← linalylacetate → 8-hydroxylinalylacetate

dihydrolinalool → 8-hydroxydihydrolinalool ← dihydrolinalylacetate → 8-hydroxydihydrolinalylacetate

Fig. 14 Transformation of linalool, dihydrolinalool and their acetates by cell cultures of *Nicotiana tabacum*

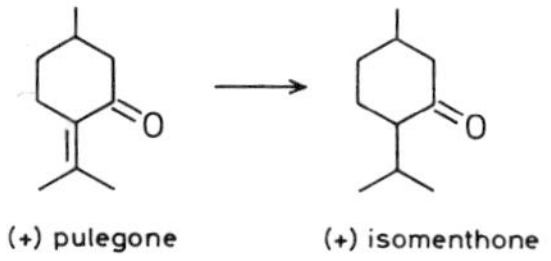

(+) pulegone → (+) isomenthone

Fig. 15 Transformation of pulegone by cell cultures of *Mentha* chemotypes

5.2 Diterpenoids

The diterpenoid stevioside is widely used as a sweetening agent. The aglycone steviol is not sweet. Biotransformation of steviol by cell cultures of *Stevia rebaudiana*

steviol → steviolbioside + stevioside

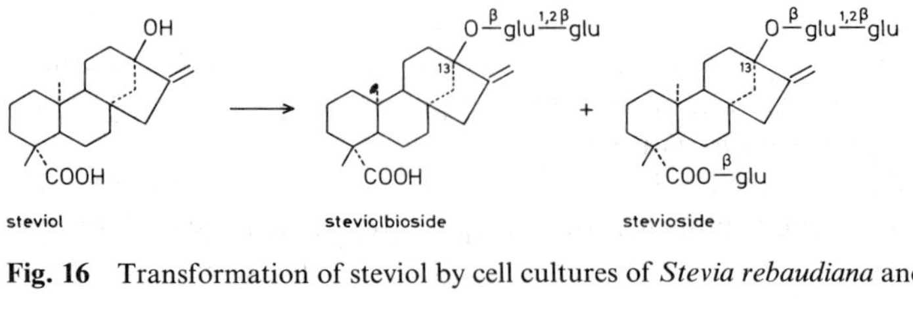

Fig. 16 Transformation of steviol by cell cultures of *Stevia rebaudiana* and *Digitalis purpurea*

and *Digitalis purpurea* has been investigated in order to obtain products with higher sweetness[35]. Steviol is glucosylated by both cell cultures to steviolbioside as well as stevioside and three compounds of unknown structure (Fig. 16).

5.3 Panaxatriol

Ginseng callus produces both as callus and as suspension culture high amounts of saponins. The saponins are thought to have a higher pharmacological activity than the genins. The aglycone panaxatriol is isolated from ginsenoside Rg-1 as an isolation artifact during technical isolation of the saponins. Therefore, panaxatriol was incubated with cell cultures of *Panax ginseng*[35]. Panaxatriol was glucosylated to panaxatriol-3-glucoside and to panaxatriol-6-glucoside as well as to some compounds whose structure has still to be elucidated (Fig. 17).

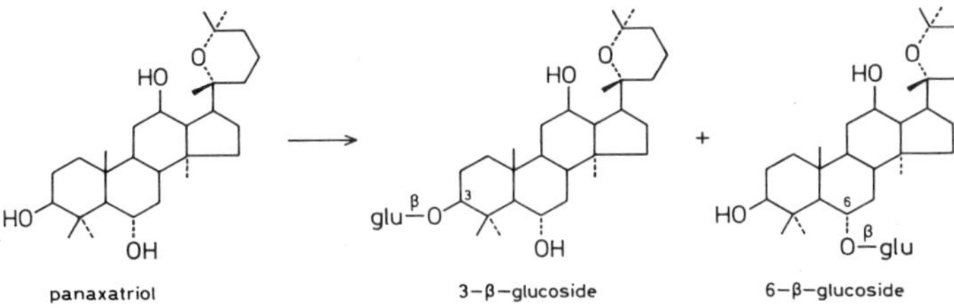

panaxatriol → 3-β-glucoside + 6-β-glucoside

Fig. 17 Transformation of panaxatriol by cell cultures of *Panax ginseng*

6 Steroids

Cell cultures offer the potential of producing rare or unusual steroids. Selected biotransformations to obtain desired products might be carried out more economically using cell cultures than by synthetic means.

6.1 Pregnane Derivatives

6.1.1 Progesterone

The transformation of progesterone (Fig. 18) has been reported for a number of cell cultures. It can be metabolized by two pathways. One involves the reduction at C-20 to give Δ^4-pregnen-20α-ol-3-one. The other involves the saturation of ring A resulting in 5α-pregnane-3,20-dione and then 5α-pregnane-3β-ol-20-one. In addition, one can observe 5α-pregnane-3β,20α-diol and 5α-pregnane-3β,20β-diol.

Some cell cultures are capable of esterification or glucosylation of transformation products of progesterone with palmitic acid or glucose, respectively (Fig. 19, 20).

These biotransformation reactions have been comprehensively covered by recent reviews[36,37] and are summarized in Table 1.

More recently, Gallili, Yagen and Mateles[38,39] reported the biotransformation of progesterone by suspension cultures of *Lycopersicum esculentum*, *Capsicum frutescens* and *Catharanthus roseus*. Cultures of *L. esculentum* and *Capsicum frutescens*

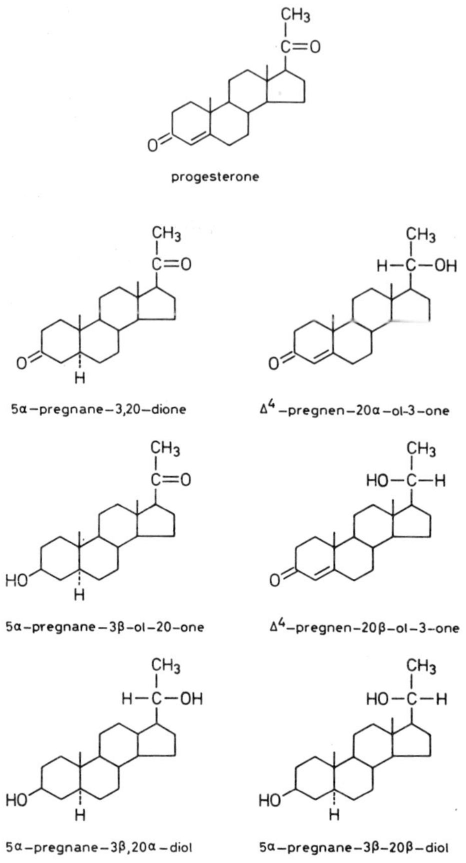

progesterone

5α-pregnane-3,20-dione

Δ^4-pregnen-20α-ol-3-one

5α-pregnane-3β-ol-20-one

Δ^4-pregnen-20β-ol-3-one

5α-pregnane-3β,20α-diol

5α-pregnane-3β-20β-diol

Fig. 18 Transformation products of progesterone after incubation with plant cell cultures

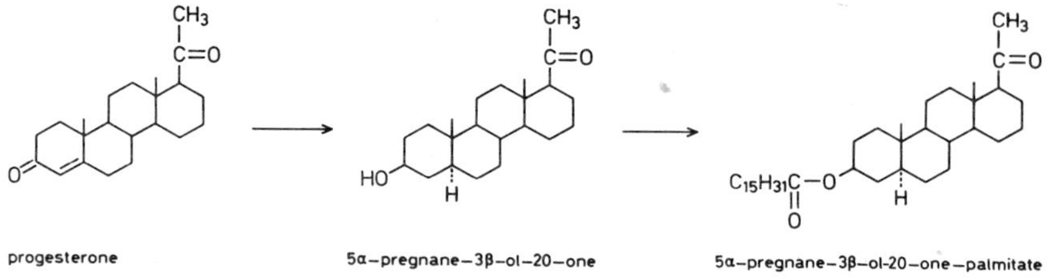

progesterone → 5α-pregnane-3β-ol-20-one → 5α-pregnane-3β-ol-20-one-palmitate

Fig. 19 Transformation of progesterone by cell cultures of *Nicotiana tabacum* and *Sophora angustifolia*

were able to reduce stereospecifically the carbonyl at C-20 to yield Δ^4-pregnen-20α-ol-3-one in contrast to cultures of *C. roseus*, which were capable of converting it into the 20β-hydroxy isomer (Fig. 18). It is most remarkable that cell cultures of *L. esculentum* and *C. roseus* could transform progesterone to Δ^4-pregnen-14α-ol-3,20-dione. This hydroxylation in the 14α-position has not been previously reported as occurring in plant cell cultures. In addition Δ^4-pregnen-6β,14α-diol-3,20-dione

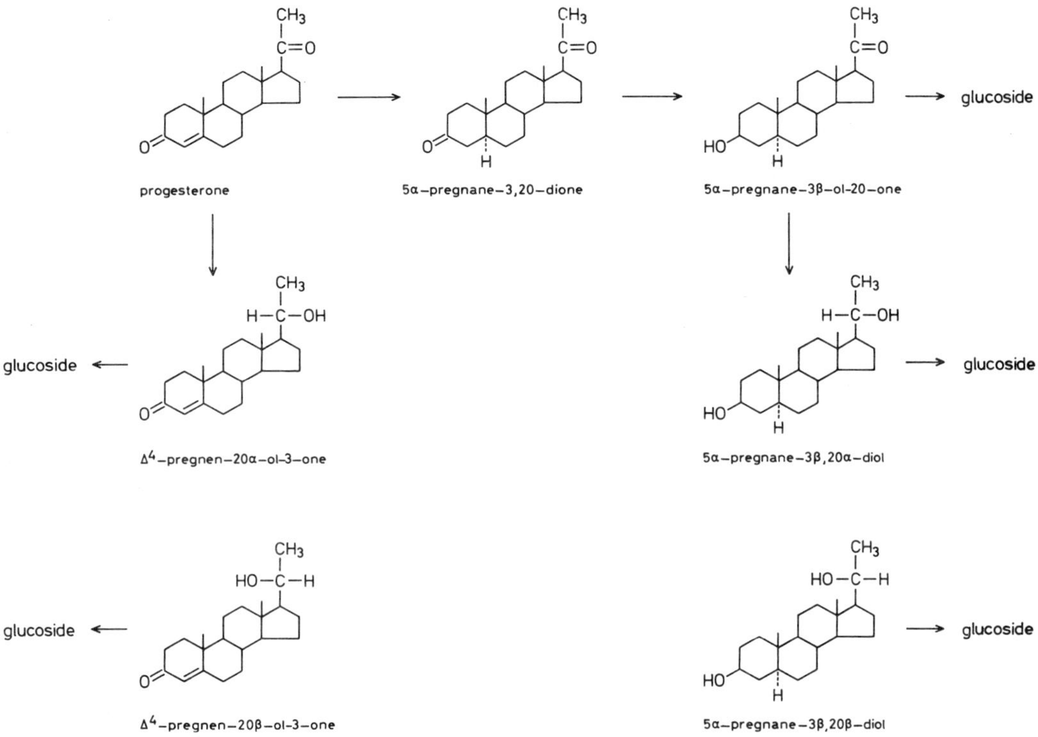

Fig. 20 Transformation of progesterone by cell cultures of *Digitalis purpurea*

and Δ^4-pregnen-6β,11α-diol-3,20-dione could be identified as new metabolites of progesterone in cell cultures of *L. esculentum* (Fig. 21).

Thus, biotransformation of progesterone by plant cell cultures includes the reduction of the Δ^4 double bond, reduction of the 3-keto and 20-keto groups, formation of palmitates and of glucosides and hydroxylation in positions 6β, 11α and 14α. In general the yields of these biotransformation products are very low compared with those obtained by microorganisms. A very notable exception are cell cultures of *Capsicum frutescens* which convert progesterone to Δ^4-pregnen-20α-ol-3-one in yields of 60–90% with relative absence of by-products[39)].

6.1.2 5β-Pregnane Derivatives

Hirotani and Furuya[40)] have reported the biotransformation of 5β-pregnane-3,20-dione to 5β-pregnane-3α,20α-diol and of 5β-pregnane-3β-ol,20-one to 5β-pregnane-3β,20α-diol by *Digitalis purpurea* suspension cultures. From all compounds the monoglucosides could be isolated as well.

In general biotransformation of 5β-pregnane derivatives is identical to that of 5α-pregnane derivatives.

Table 1. Transformation reactions of the progesterone molecule by plant cell cultures

Plant species and (literature reference)	Reduction of the Δ^4-double bond	3-β-Reduction	20-α-Reduction	20-β-Reduction	Palmitate	Glucoside	6-β-Hydroxylation	11-α-Hydroxylation	14-α-Hydroxylation
Digitalis purpurea[35]	+	+	+	+		+			
Digitalis purpurea[41]	+	+							
Sophora angustifolia[42]	+	+			+				
Nicotiana tabacum[42] Bright Yellow	+	+			+				
Dioscorea deltoidea[37]	+	+		+		+			
*Dioscorea deltoidea**[37]	+								
Cheiranthus cheiri[37]	+								
Parthenocissus spec.[41]			+						
Rosa spec.[41]	+	+	+	+					
Solanum tuberosum[41]	+	+							
Digitalis lutea[41]	+								
Atropa belladonna[41]	+	+							
Nicotiana rustica[41]	+	+							
Hedera helix[41]	+	+							
Catharanthus roseus[38]				+					+
Capsicum frutescens[39]			+						
Lycopersicum esculentum[39]	+	+	+				+	+	+

* Microsome fraction

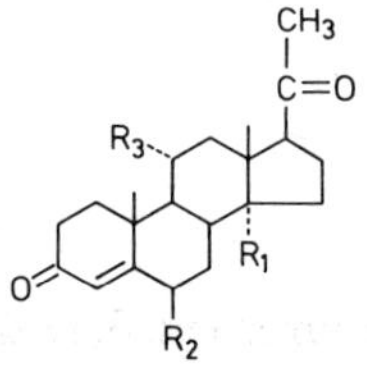

CH_3 C=O R_3 R_1 O R_2

R_1	R_2	R_3	
OH	H	H	14α-hydroxyprogesterone (Δ^4-pregnene-14α-ol-3,20-dione)
OH	OH	H	6β,14α-dihydroxyprogesterone (Δ^4-pregnene-6β,14α-diol-3,20-dione)
H	OH	OH	6β,11α-dihydroxyprogesterone (Δ^4-pregnene-6β,11α-diol-3,20-dione)

Fig. 21 Hydroxylation products of progesterone by suspension cultures of *Catharanthus roseus* (14α), and *Lycopersicum* esculentum (6β, 11α, 14α)

6.1.3 Pregnenolone

Graves and Smith[41)] reported that pregnenolone (Fig. 22) is converted to progesterone by cell cultures of *Digitalis purpurea*, *D. lutea* and *Nicotiana tabacum*. This transformation involves oxidation of the hydroxy group at position 3 and a shift of the Δ^5 double bond to the Δ^4 position.

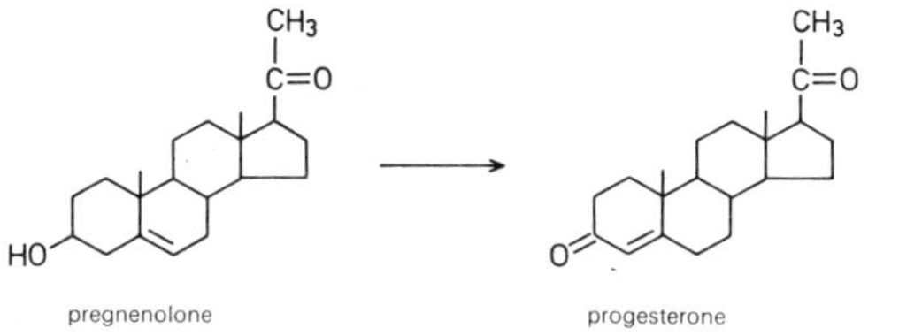

Fig. 22 Transformation of pregnenolone by cell cultures of *Digitalis purpurea*, *D. lutea* and *Nicotiana tabacum*

Furuya *et al.*[42)] demonstrated that pregnenolone (Fig. 23) was transformed to pregnenolone palmitate and 5α-pregnanolone palmitate by suspension cultures of *N. tabacum* var. Bright Yellow and of *Sophora angustifolia*.

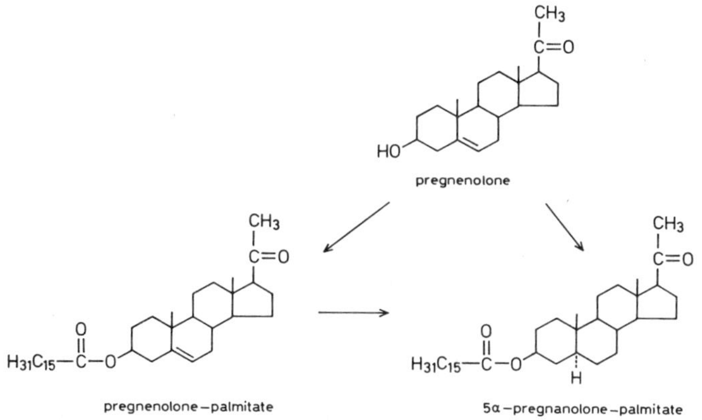

Fig. 23 Transformation of pregnenolone by cell cultures of *Sophora angustifolia* and *Nicotiana tabacum* (Bright Yellow)

The transformation reactions reported for pregnenolone are summarized in Table 2.

6.2 Androstene Derivatives

Stohs and El-Olemy[43)] studied the biotransformation of Δ^4androstene-3,17-dione to 5α-androstane-3β-ol-17-one and 5α-androstane-3β,17β-diol by suspension cultures of *Dioscorea deltoidea*. These metabolites were present in a conjugated form, probably as glycosides.

Table 2. Transformation reactions of the pregnenolone molecule by plant cell cultures

Plant species (reference)	3-oxidation	Shift of Δ^5 to Δ^4 double bond	Reduction of the Δ^4 double bond	Reduction of Δ^5 double bond	Palmitate formation
Digitalis purpurea[41)]	+	+			
Digitalis lutea[41)]	+	+	+		
Nicotiana tabacum[41)]	+	+			
Nicotiana tabacum[42)] Bright Yellow	+	+	+	+?	+
Sophora angustifolia[42)]	+	+	+	+?	+

Hirotani and Furuya[44)] have examined the biotransformation of testosterone by cell cultures of *N. tabacum* var. Bright Yellow. As conversion products, they found Δ^4-androstene-3,17-dione, 5α-androstane-3β-ol-17-one and its palmitate, 5α-androstane-17β-ol-3-one and 5α-androstane-3β,17β-diol and its palmitate (Fig. 24).

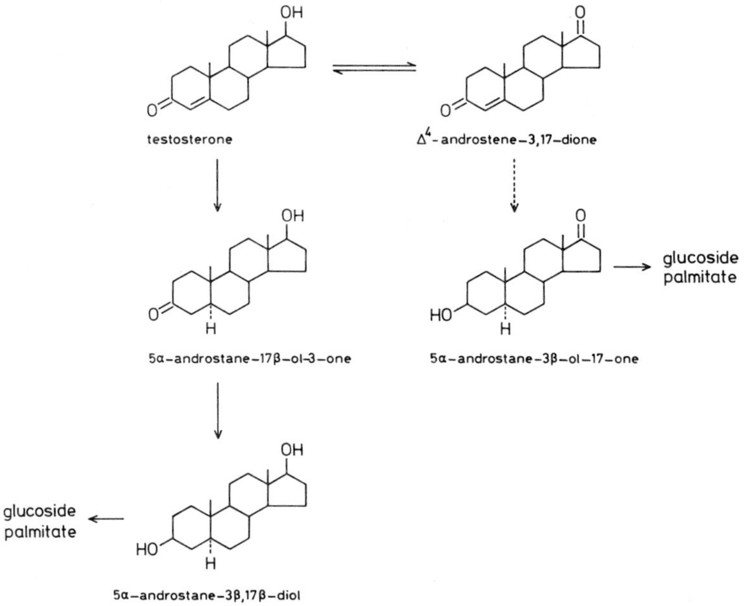

Fig. 24 Transformation of testosterone by cell cultures of *Nicotiana tabacum* (Bright Yellow)

These metabolic products are almost the same as observed in mammals, except that the formation of palmitates and glucosides is very distinctive.

The biotransformation reactions of the androstene molecule are summarized in Table 3.

Table 3. Transformation reactions of the Δ^4 androstene molecule by plant cell cultures

Plant species (reference)	17-Oxidation	Reduction of the Δ^4 double	17-β-Reduction	3-β-Reduction	Palmitate formation	Glucoside formation	Conjugated form
Dioscorea deltoidea[43]		+	+	+			+
Nicotiana tabacum var. Bright Yellow[44]	+	+	+	+	+	+	

6.3 Cholesterol

In reactions analogous to those seen with pregnanes and androgens, cholesterol can be transformed by several plant cell cultures. Stohs and El-Olemy[45] found that homogenates of *Cheiranthus cheiri* cell cultures are capable of metabolizing cholesterol to Δ^4-cholesten-3-one. This compound can be further metabolized to 5α-cholestan-3-one by homogenates of cell cultures of *Digitalis purpurea*, *Cheiranthus cheiri*, *Strophanthus sarmentosus* and *Dioscorea deltoidea*[37].

Weber[46] reported the conversion of cholesterol to 5α-cholestan-3β-ol by *Brassica napus* and *Glycine max* suspension cultures (Fig. 25).

HO cholesterol → HO H 5α-cholestan-3β-ol

Fig. 25 Transformation of cholesterol to 5α-cholestan-3β-ol by suspension cultures of *Brassica napus* and *Glycine max*

6.4 Cardenolides

Cardiac glycosides used in medicine have still to be isolated from plant material. Their chemical synthesis is still not applicable on a technical scale. In the USA, digoxin and digitoxin rank sixth and eighth among compounds of plant origine used in medicine and together make up one percent of all prescriptions[47]. Because of its better pharmacokinetic behaviour, the use of digoxin is increasing at the expense of digitoxin. *Digitalis lanata* plants, the source of digoxin, always contain high amounts of digitoxin, which has to be separated from digoxin during the isolation. In the factories, increasing amounts of digitoxin are stored. Digoxin differs from digitoxin only by an additional hydroxyl function at C-12 (Fig. 26).

Already in the sixties biotransformation of cardenolides by microorganisms had been investigated[48, 49, 50]. Microorganisms can hydroxylate the aglycones, whereas C-12 hydroxylation of glycosides does not occur or is incomplete[51]. Also hydroxylation by chemical means is not applicable. Therefore especially, biotransformation of

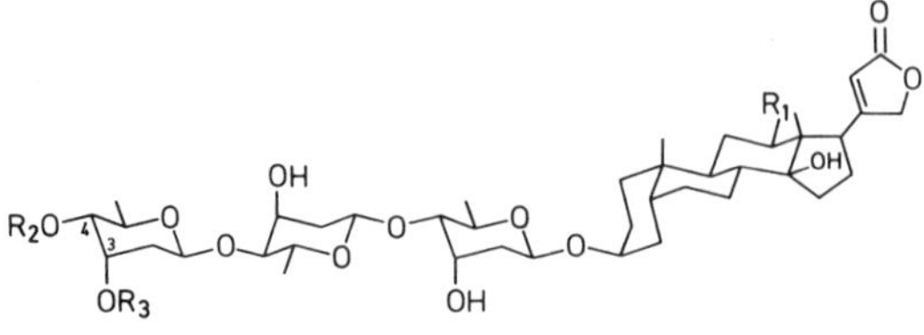

Fig. 26

	R^1	R^2	R^3
Digitoxin	H	H	H
β-Methyldigitoxin	H	-methyl	H
Purpurea glycoside A	H	-glucose	H
Lanatoside A	H	-glucose	-acetyl
Digoxin	OH	H	H
β-Methyldigoxin	OH	-methyl	H
Deacetyllanatoside C	OH	-glucose	H
Lanatoside C	OH	-glucose	-acetyl

cardiac glycosides by plant cell cultures has been investigated very intensively in recent years.

From a pharmacological point of view the cardiac glycosides used in therapy are far from being very satisfactory. The therapeutic range is narrow, additionally there is a wide difference in the response of the patients to these compounds. Therefore, it would be very important to find compounds in which the positive inotropic activity is not connected with the toxic property. Many chemical derivatives of cardiac glycosides have been made by chemical modifications. However, up to now "none of these modifications have resulted in a compound either with pharmacological properties superior to the classical glycosides or with better therapeutic index"[52]. It is possible to produce compounds not yet found in nature by means of biotransformation with plant cell cultures, hence, this method is of interest in the field of cardenolides, too. Furthermore, the results presented later on can demonstrate the strategy used to increase productivity of secondary product formation by plant cell culture techniques.

6.4.1 Biotransformation Patterns

6.4.1.1 Aglycones

Biotransformation of digitoxigenin (Fig. 27) has been achieved with cell cultures of plants containing cardenolides[35, 53–56] as well as with those of plants containing no cardenolides[57, 58]. Digitoxigenin is hydroxylated at C-12 by cell cultures of *D. lanata* and after addition of digitoxose formation of the digitoxoside is observed[55]. Without addition of digitoxose only glucosylation occurs[35, 56] as well as oxidation to digitoxigenone[54]. Formation of digitoxigeninglucoside and of digitoxigenone has also been observed with *Thevetia*[56] and *Digitalis purpurea* cell cultures[35]. In the latter case, additional formation of epidigitoxigenin, epidigitoxigenin-β-D-glucoside, 5-β-hydroxy- and 7-β-hydroxydigitoxigenin has been demonstrated[35]. Very rapid biotransfor-

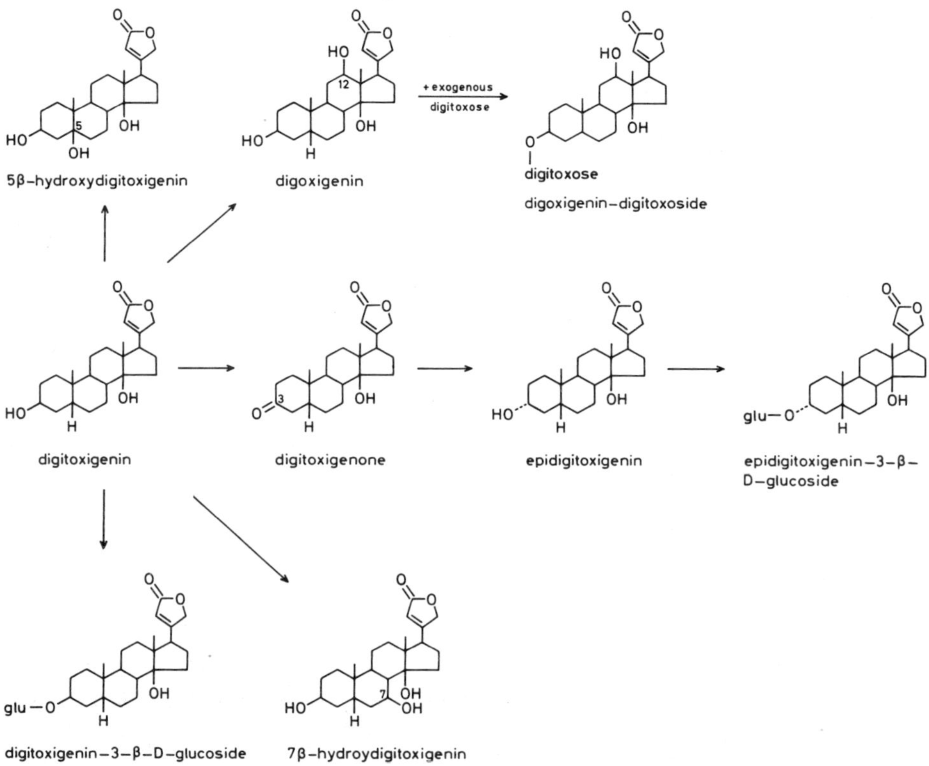

5β-hydroxydigitoxigenin

digoxigenin

+ exogenous digitoxose

digitoxose

digoxigenin-digitoxoside

digitoxigenin

digitoxigenone

epidigitoxigenin

epidigitoxigenin-3-β-D-glucoside

digitoxigenin-3-β-D-glucoside

7β-hydroydigitoxigenin

Fig. 27 Transformation of digitoxigenin by cell cultures of *Digitalis*

mation of digitoxigenin by cell cultures of *Cannabis sativa*, *Ipomoea sp.*, and *Daucus carota* strain Ca 68 was reported by Veliky and coworkers[57, 58]. The product was identified as 5β-hydroxydigitoxigenin (= periglogenin), and in submerse cultures of a working volume of 5 l with pH regulated at 4.8 ± 0.2 up to about 50 mg l^{-1} of periplogenin could be produced during 48 h. Thus, biotransformation patterns of digitoxigenin with plant cell cultures are quite similar to those obtained with microorganisms, except for glycoside formation[48-50].

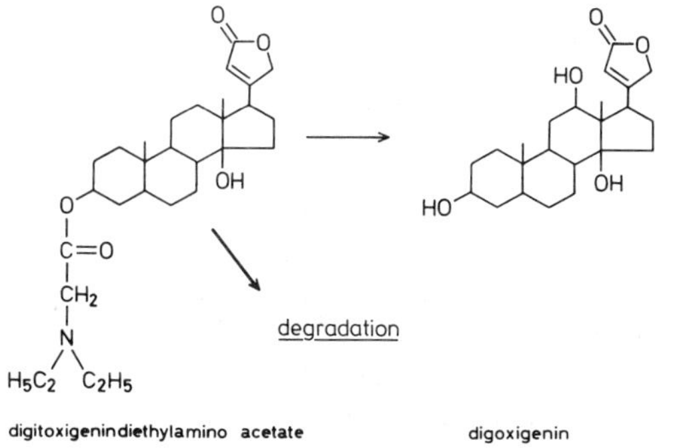

degradation

digitoxigenindiethylamino acetate

digoxigenin

Fig. 28 Transformation of digitoxigenin-diethylaminoacetate by cell cultures of *Digitalis lanata*

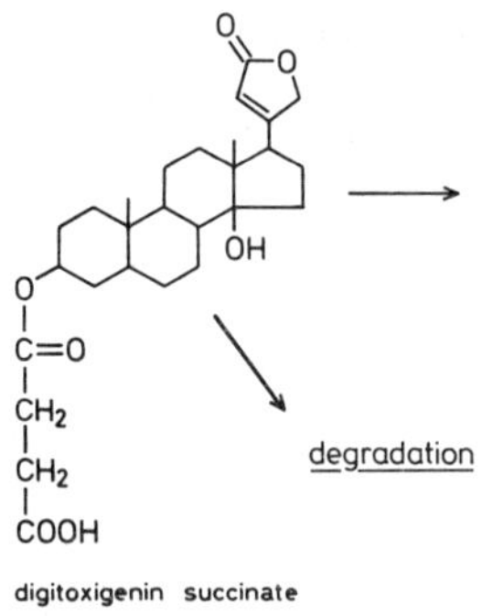

Fig. 29 Transformation of digitoxigenin-succinate by cell cultures of *Digitalis lanata*

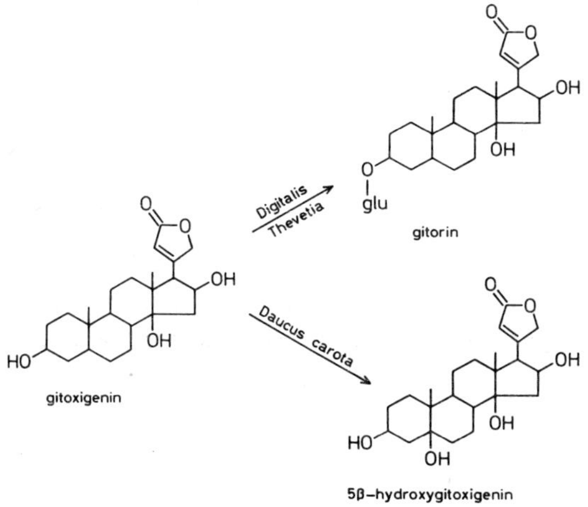

Fig. 30 Transformation of gitoxigenin by cell cultures of *Digitalis lanata*, *Thevetia neriifolia* and *Daucus carota*

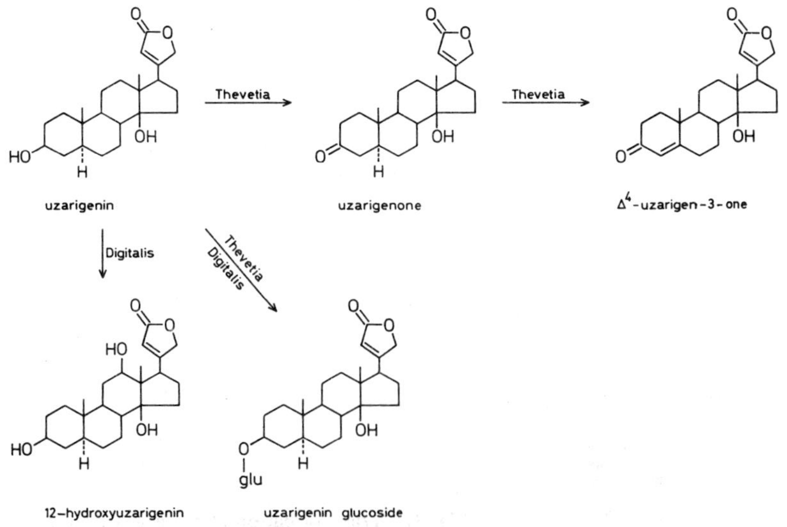

Fig. 31 Transformation of uzarigenin by cell cultures of *Digitalis lanata* and *Thevetia neriifolia*

Digitoxigenin-diethylaminoacetate (Fig. 28) as well as digitoxigenin-succinate (Fig. 29) are hydrolyzed by cell cultures of *D. lanata*, and formation of low amounts of digoxigenin occurs. The main reaction, however, is a degradation of the substrate[55].

Gitoxigenin (Fig. 30) is glucosylated by cell cultures of *D. lanata* as well as *Thevetia* to gitorin. Two further products obtained with *Thevetia* were not yet identified[56]. Cell cultures of *Daucus* hydroxylate gitoxigenin to 5-β-hydroxygitoxigenin, a compound not yet found in nature[58].

Uzarigenin (Fig. 31) differs from digitoxigenin only by the fact that the rings A and B of the steroid skeleton are linked trans instead of cis to each other. *Thevetia* cell cultures were found to be able to transform uzarigenin into uzarigenone, Δ^4-uzarigen-3-one and uzarigeninglucoside. *Digitalis* cell cultures produced uzarigeninglucoside and a compound presumably hydroxylated at C-12, a compound up to now unknown in nature[56].

6.4.1.2 Monoglycosides

Biotransformation of cardiac monoglycosides has been investigated recently[56, 59, 60]. Digitoxigenindigitoxoside (Fig. 32) is glucosylated to glucodigitoxigenindigitoxoside by *Digitalis leucophaea* strain 30614-9-3S and *D. parviflora*[60]. Glucosylation of digoxigenin- as well as of gitoxigenindigitoxoside (Fig. 32) by cell cultures of *D. lanata* was observed, too[56]. Furtheron, cymarin (Fig. 33) was glucosylated to k-strophanthin-β by cell cultures of *Thevetia neriifolia*[59]. After incubation of peruvoside (Fig. 34) with *D. lanata*, formation of "diglycoside A", theveneriin and neriifolin was

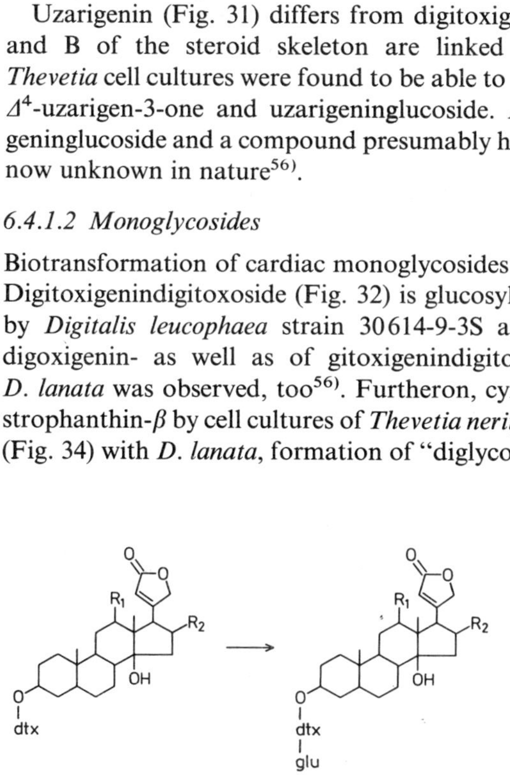

	R_1	R_2	
digitoxigenin digitoxoside	H	H	glucodigitoxigenin digitoxoside
digoxigenin digitoxoside	OH	H	glucodigoxigenin digitoxoside
gitoxigenin digitoxoside	H	OH	glucogitoxigenin digitoxoside

Fig. 32 Transformation of cardiac monoglycosides by cell cultures of *Digitalis lanata* and *D. leucophaea*

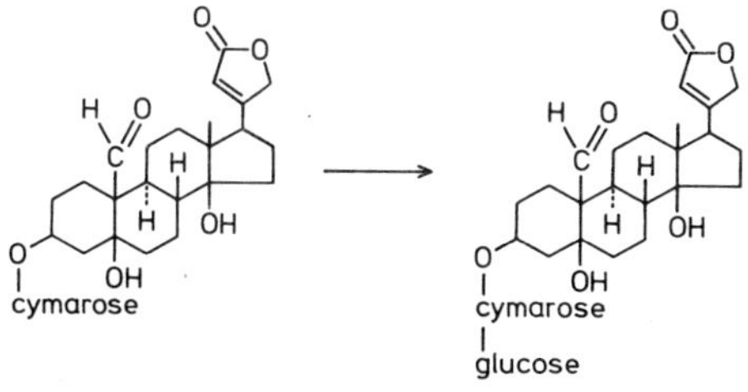

Fig. 33 Transformation of cymarin by cell cultures of *Thevetia neriifolia*

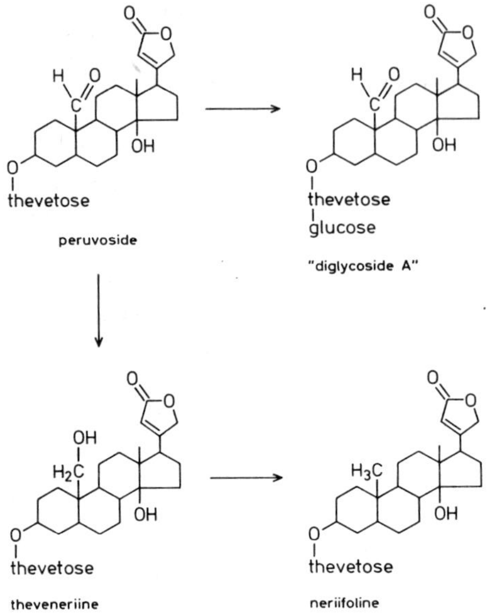

peruvoside

"diglycoside A"

theveneriine

neriifoline

Fig. 34 Transformation of peruvoside by cell cultures of *Digitalis lanata* and *Thevetia neriifolia*

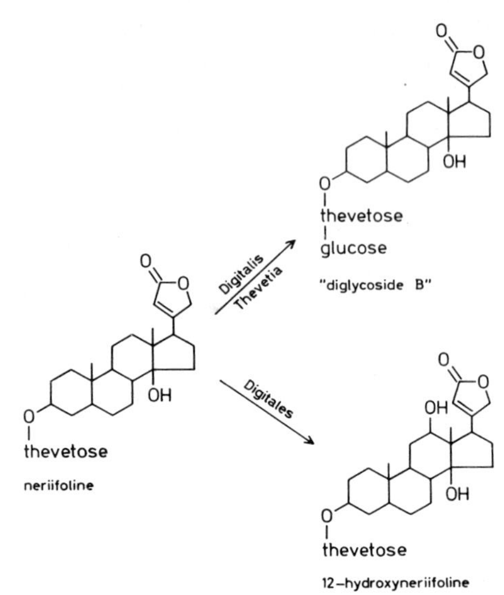

Fig. 35 Transformation of neriifoline by cell cultures of *Digitalis lanata* and *Thevetia neriifolia*

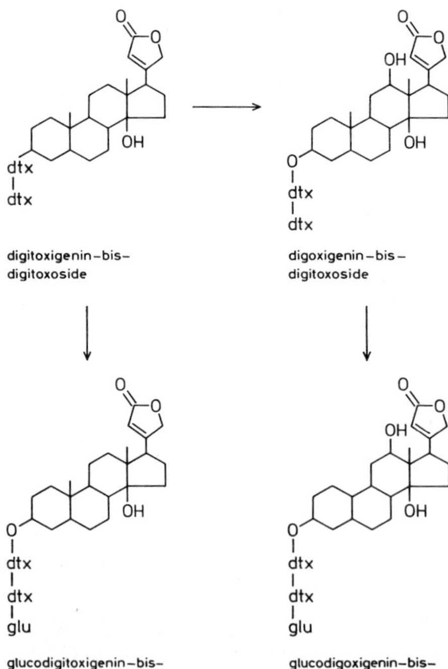

Fig. 36 Transformation of digitoxigenin-bis-digitoxoside by cell cultures of *Digitalis lanata* and *D. leucophaea*

observed, and with *Thevetia* the same result except for theveneriin formation[59]. After incubation of neriifolin (Fig. 35), glucosylation to the "diglycoside B" occurred with cells of *D. lanata* and *Thevetia*, in the experiments with *D. lanata* a compound was found which was presumably neriifolin hydroxylated at C-12, a substance up to now not found in nature[56].

6.4.1.3 Diglycosides

Digitoxigenin-bis-digitoxoside (Fig. 36) is glucosylated to the appropriate glycoside by cell cultures of *D. lanata*[56] as well as of *D. leucophaea* and *D. cariensis ssp. trojana*[60].

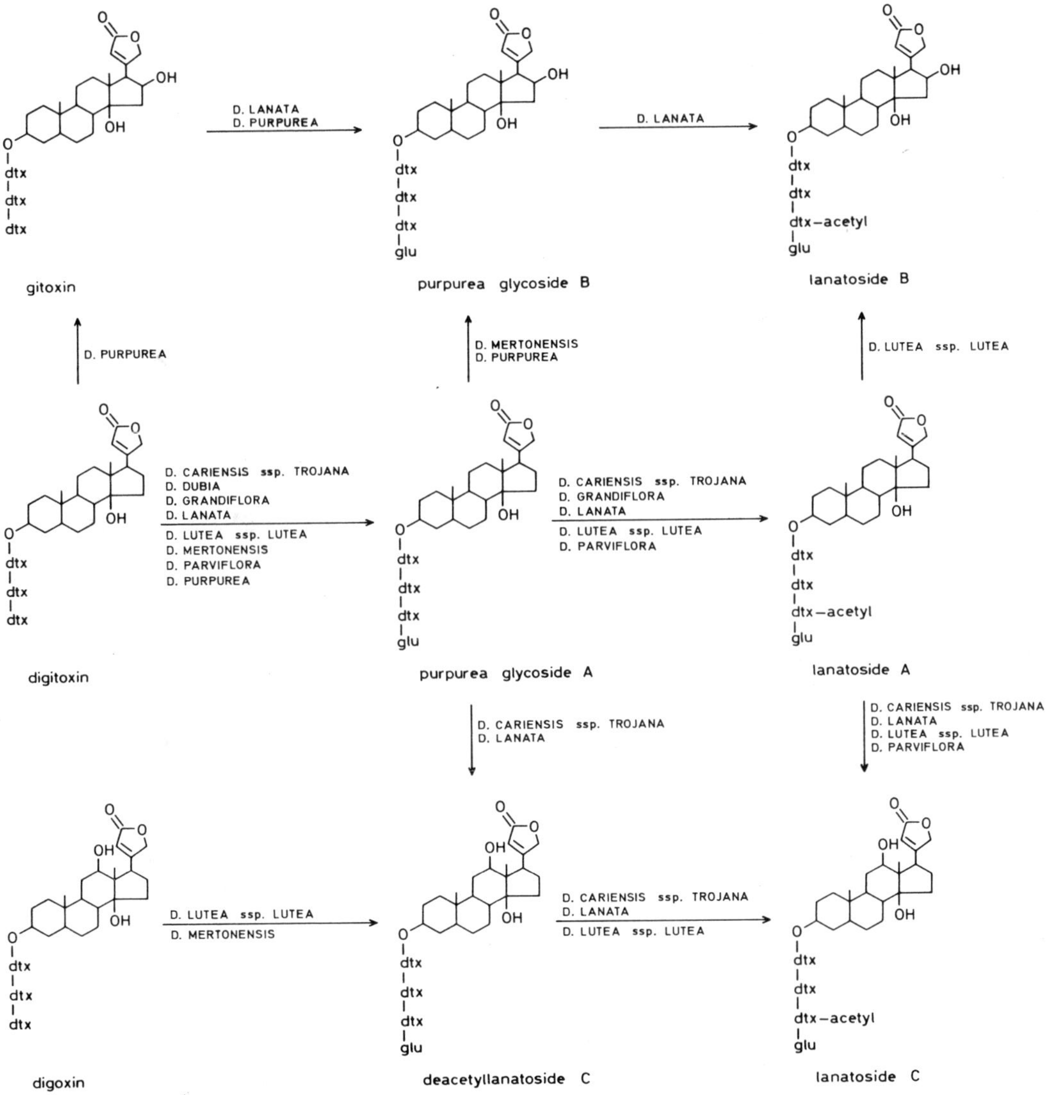

gitoxin (O–dtx–dtx–dtx)
D. LANATA
D. PURPUREA
purpurea glycoside B (O–dtx–dtx–dtx–glu)
D. LANATA
lanatoside B (O–dtx–dtx–dtx-acetyl–glu)

D. PURPUREA
D. MERTONENSIS
D. PURPUREA
D. LUTEA ssp. LUTEA

digitoxin (O–dtx–dtx–dtx)
D. CARIENSIS ssp. TROJANA
D. DUBIA
D. GRANDIFLORA
D. LANATA
D. LUTEA ssp. LUTEA
D. MERTONENSIS
D. PARVIFLORA
D. PURPUREA
purpurea glycoside A (O–dtx–dtx–dtx–glu)
D. CARIENSIS ssp. TROJANA
D. GRANDIFLORA
D. LANATA
D. LUTEA ssp. LUTEA
D. PARVIFLORA
lanatoside A (O–dtx–dtx–dtx-acetyl–glu)

D. CARIENSIS ssp. TROJANA
D. LANATA
D. CARIENSIS ssp. TROJANA
D. LANATA
D. LUTEA ssp. LUTEA
D. PARVIFLORA

digoxin (O–dtx–dtx–dtx)
D. LUTEA ssp. LUTEA
D. MERTONENSIS
deacetyllanatoside C (O–dtx–dtx–dtx–glu)
D. CARIENSIS ssp. TROJANA
D. LANATA
D. LUTEA ssp. LUTEA
lanatoside C (O–dtx–dtx–dtx-acetyl–glu)

Fig. 37 Transformation of digitoxin, digoxin and gitoxin by cell cultures of different *Digitalis sp.*

In the experiments with *D. lanata* formation of digoxigenin-bis-digitoxoside has been observed, which when fed to the cells can also be glucosylated[56]. In none of these experiments biosynthesis of digitoxose has been detected up to now.

6.4.1.4 Triglycosides

Biotransformation of cardiac triglycosides by cell cultures of *Digitalis* and *Thevetia* species has been investigated intensively by Furuya as well as Reinhard and coworkers[36, 55, 56, 59–63]. Fig. 37 summarizes the different biotransformation patterns found using digitoxin, digoxin and gitoxin as substrates. In all cases investigated these "secondary glycosides" are glucosylated. *Digitalis mertonensis* and *D. purpurea* are

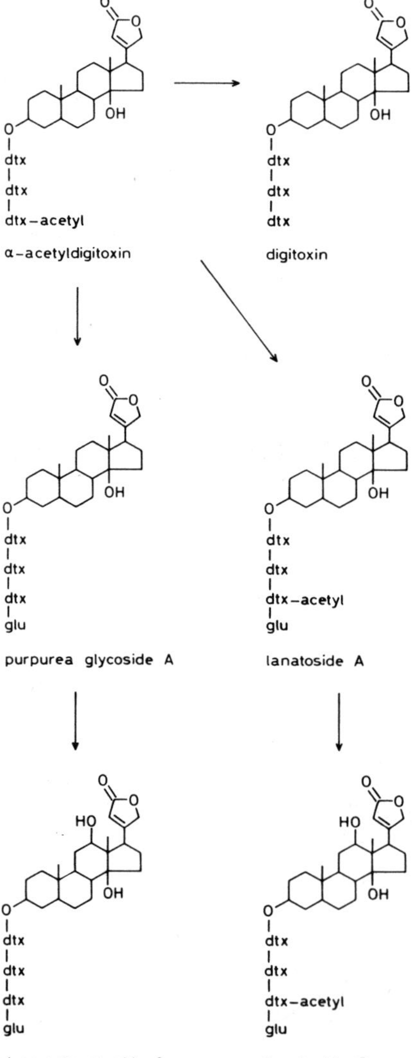

Fig. 38 Transformation of α-acetyldigitoxin by cell cultures of *Digitalis lanata*

able to hydroxylate digitoxin and purpurea glycoside A (PA) at C-16 to gitoxin and purpurea glycoside B (PB), respectively. Cell cultures initiated from plants containing lanatosides like *D. cariensis ssp. trojana*, *D. grandiflora*, *D. lanata*, *D. lutea ssp. lutea* and *D. parviflora* acetylate PA, PB or deacetyllanatoside C (DC) to the corresponding lanatosides. Of these cell cultures, those from plants containing the so-called C-glycosides hydroxylate the "A-glycosides" at C-12, and DC and lanatoside C can be found along with digoxin, which may occur by direct hydroxylation of digitoxin or by deacetylation and deglucosylation of LC and DC. The amounts of the single products differ significantly from cell strain to cell strain. Cell cultures of *D. lutea* also hydroxylate lanatoside A at C-16 to lanatoside B. This reaction was not yet found with cells of *D. lanata* or *D. leucophaea*, presumably, because these cell strains have been isolated from plants containing only low amounts of B-glycosides. Gitoxin, however, is as to be expected glucosylated and acetylated by cell cultures of *D. lanata* to purpurea glycoside B and lanatoside B. On the other hand, cell cultures of *D. mertonensis* are also able to glucosylate digoxin to DC, a reaction which does not occur in the plant, because this plant is unable to produce the precursor digoxin. α-Acetyldigitoxin (Fig. 38) is glucosylated by cell cultures of *D. lanata* to lanatoside A, which is hydroxylated to lanatoside C. The substrate is simultaneously deacetylated to digitoxin. This is then glucosylated to PA, which can be hydroxylated to DC[36, 63]. After application of digitoxindisuccinate to these cell cultures the succinoyl residues are split off during the first seven days of incubation. Afterwards, the same biotransformation products are obtained as with digitoxin[55].

Thevetin A and B (Fig. 39) are deglucosylated by cell cultures of both *Digitalis lanata* and *Thevetia neriifolia* to peruvoside and neriifoline, respectively[59].

6.4.1.5 Tetraglycosides

Lanatoside A (Fig. 40) is metabolized rapidly by cell cultures of *Digitalis lanata*, whereby the biotransformation patterns found depend on the cell strain and the cul-

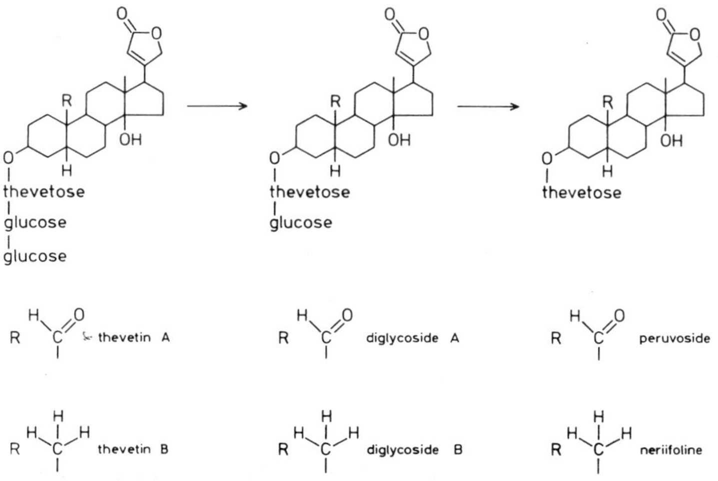

thevetose – glucose – glucose → thevetose – glucose → thevetose

R = CHO: thevetin A → diglycoside A → peruvoside

R = CH₃: thevetin B → diglycoside B → neriifoline

Fig. 39 Transformation of thevetin A and B by cell cultures of *Thevetia neriifolia* and *Digitalis lanata*

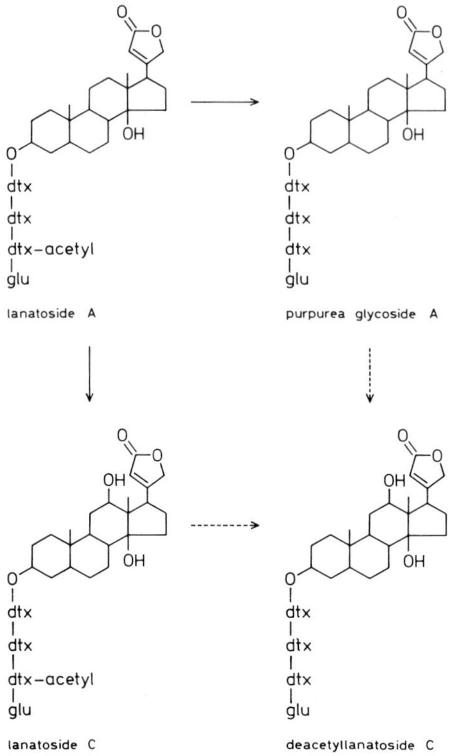

Fig. 40 Transformation of lanatoside A by cell cultures of *Digitalis lanata*

tural conditions. Under light conditions strain 72L produces LC, PA, and DC, whereas with strain 72D (dark conditions) only formation of LC, and no deacetylation can be observed[36,63]. In experiments with *Thevetia* only deacetylation and deglucosylation to digitoxin occurred[59]; the same reaction was found with LB and LC in formation of gitoxin and digoxin[56]. After application of LA (Fig. 41), cell cultures of *D. lutea* produced α-acetyldigitoxin and α-acetyldigoxin besides lanatoside C[55].

6.4.2 Screening of Digitalis Cell Strains for Hydroxylating Capacity

As already mentioned above, C-12 hydroxylation of the A-glycosides is the most important reaction from a medicinal point of view. Use of this reaction on a biotechnological scale would demand fulfilling the prerequisite that the administered substrate is transformed quickly, totally and to a single product. Digitoxin presents itself as a substrate, because it occurs during technical isolation of digoxin as a by-product. As shown earlier, digitoxin, however, is transformed into different products, and especially C-12 hydroxylation is very incomplete. Therefore, biotransformation of different digitoxin derivatives was investigated, and first experiments showed that β-methyldigitoxin was transformed quantitatively to β-methyldigoxin. β-Methyldigoxin is already used in medicine. Up to now it is made by chemical methylation of digoxin[36]. Using shake cultures a broad screening of cell strains has been undertaken in order to find cell strains with better hydroxylation capacities. These cell cultures

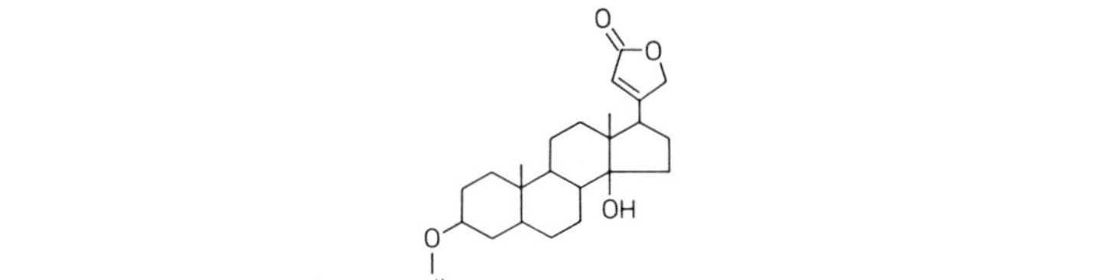

α-acetyldigitoxin

lanatoside A — lanatoside C — α-acetyldigoxin

Fig. 41 Transformation of lanatoside A by cell cultures of *Digitalis lutea*

had been initiated from plants with high C-glycoside content. However, not all cell cultures turned out to be only hydroxylating ones. Three types of strains have been found up to now[55, 64].

6.4.2.1 Demethylation, Glucosylation (Fig. 42, Table 4)

A first group of cell strains (128, 293 BHL, 447, 445, 130, 228, 508 and 216) cannot hydroxylate *β*-methyldigitoxin at C-12. Only demethylation and glucosylation of *β*-

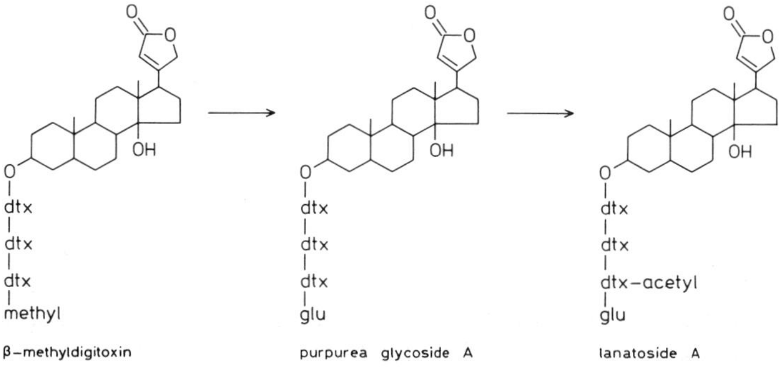

β-methyldigitoxin — purpurea glycoside A — lanatoside A

Fig. 42 Biotransformation of *β*-methyldigitoxin by cell strains 128, 293BHL, 447, 445, 130, 228, 508, 216 of *Digitalis* lanata. Lanatoside A was only found in experiments with strain 447

Table 4. Biotransformation products formed after incubation of β-methyldigitoxin with cell cultures of *D. lanata* and *D. leucophaea*. Shake cultures with 40 mg l^{-1} substrate (Data from 55 and 64)

Strain	Mdg	DC	LC	PA	LA	Mdt (rest)
291	++++	—	—	—	—	—
287	++++	—	—	—	—	—
72D	++++	—	—	—	—	—
30625-10-15S	++++	—	—	—	—	—
285	++++	—	—	(+)	—	—
9-14	++++	—	—	+	—	—
10-2	++++	—	—	+	—	—
347	++++	—	—	+	—	—
B2	++++	—	—	+	—	—
72L	+++	—	—	++	—	—
293	+++	+	—	++	—	—
23-17	+++	—	—	+	—	+
364	++	—	—	++	—	++
E1	++	—	(+)	+++	—	—
E2	+	—	—	++++	—	—
22-6	(+)	—	—	(+)	—	++++
128	—	—	—	+++	—	+
293BHL	—	—	—	+++	—	+
447	—	—	—	+++	(+)	++
445	—	—	—	+++	—	++
130	—	—	—	+++	—	++
228	—	—	—	+	—	++++
508	—	—	—	+	—	++++
216	—	—	—	(+)	—	++++

(+) 10%
\+ 10–29%
++ 30–49%
+++ 50–79%
++++ 80–100% of the reaction mixture

methyldigitoxin leading to the production of purpurea glycoside A is observed. In the case of strain 447 also trace amounts of lanatoside A are formed. In the experiments with cell strain 216, only traces of PA could be detected. Nearly all of the added substrate remained untransformed[64].

6.4.2.2 C-12 Hydroxylation Simultaneously Connected with Demethylation, Glucosylation and Acetylation (Fig. 43, Table 4)

The cell strains 9-14, 10-2, 347, B2, 72L, 293, 23-17, 364, E1, E2, and 22-6 of *D. lanata* represent a second group which is able to hydroxylate the substrate at C-12. Simultaneously, however, these strains demethylate β-methyldigitoxin and glucosylate it to purpurea glycoside A. On the one hand the main reaction product may be PA (i.e. with strain E2), on the other hand β-methyldigoxin (i.e. with strain 9-14). Small amounts of LC or DC may also be formed[64].

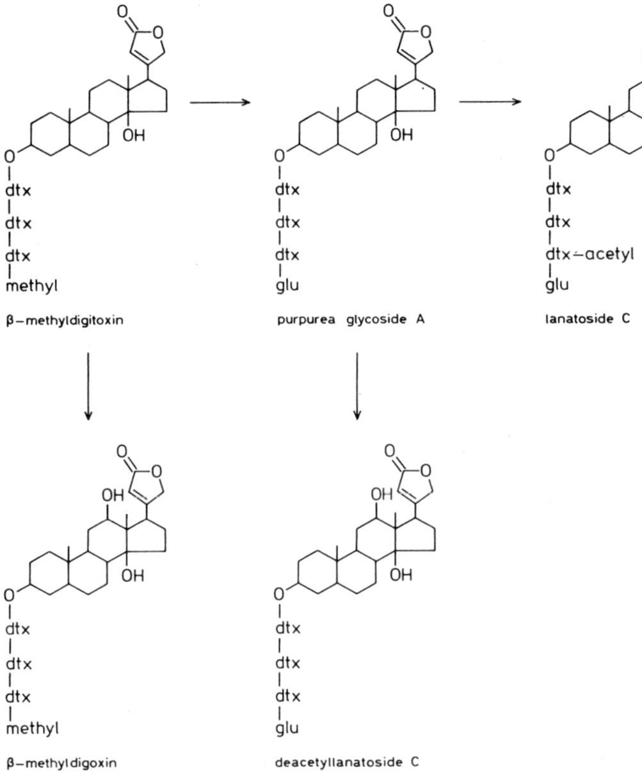

Fig. 43 Biotransformation of β-methyldigitoxin by cell strains 9-14, 10-2, 347, B2, 72L, 293, 23-17, 364, E1, E2, 22-6 of *Digitalis lanata*. Deacetallanatoside C was formed only by strain 293, lanatoside C only by strain E1

6.4.2.3 Selective C-12 Hydroxylation (Fig. 44, Table 4)

A third group of cell strains — strain 30625-10-15S of *D. leucophaea* and strains 291, 287, 285 and 72D of *D. lanata* hydroxylate the substrate almost quantitatively to β-methyldigoxin. Such cell strains were selected for elaborating a technical process of biotransformation of A- to C-glycosides[55, 64].

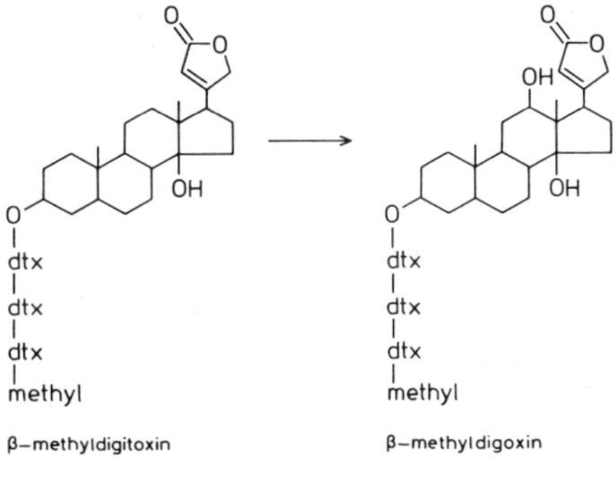

Fig. 44 Transformation of β-methyldigitoxin by cell strains 291, 287, 72D, 285 of *Digitalis lanata* and 30625-10-15S of D. leucophaea. In addition, strain 285 produces trace amounts of purpurea glycoside A

6.4.2.4 C-16 Hydroxylation (Fig. 45)

As already mentioned, in none of the experiments with cell cultures of *D. leucophaea* or *D. lanata* hydroxylation of an A-glycoside at C-16 could be observed, in contrast to experiments with *D. purpurea*[61,62], *D. lutea* or *D. mertonensis*[55,60]. After incubation of β-methyldigitoxin, cell cultures of *D. lutea* as well as of *D. mertonensis* hydroxylated part of the substrate at C-16 to β-methylgitoxin[55,60].

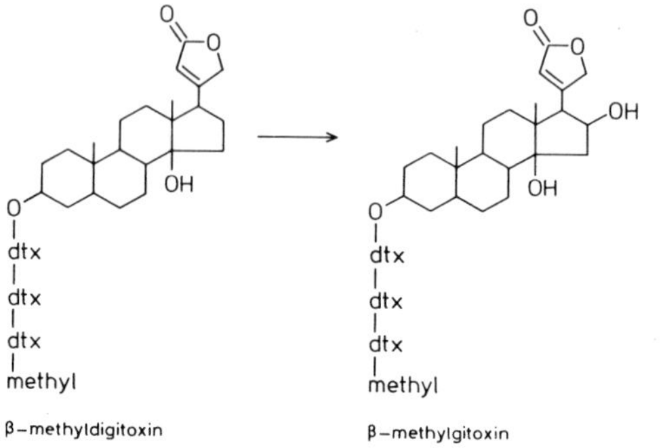

β-methyldigitoxin → β-methylgitoxin

Fig. 45 C-16 hydroxylation of β-methyldigitoxin by cell cultures of *Digitalis mertonensis* and *D. lutea*

6.4.3 Biotransformation of β-Methyldigitoxin in Bioreactors

In order to use biotransformation reactions by plant cells on a technical scale it is necessary to reproduce results obtained in shake cultures in bioreactors. Since the pioneering experiments of Tulecke and Nickel[65], only a few papers were published about cultivation of plant cells in reactors with working volumes of more than 10 l (for reviews see[66,67]). In most cases normal bioreactors with flat turbine impellers were used (i.e.[68–72]), but also bubble columns[68] and more recently such with Kaplan turbine or airlift reactors[73,74]. *Digitalis* cell cultures turned out to be sensitive to stirring and no biotransformation could be achieved using normal reactors with blade stirrers or a Kaplan turbine[75]. The airlift reactor is characterized by low shear stress, and it was shown that secondary product formation by *Morinda* cell cultures was higher in this reactor than in other types[73,74]. Therefore, further experiments were done in airlift reactors with a working volume of 20 l[55,75]. Up to now the cell strains 72L, E1, 291, 347, and 287 (Table 4) were tested. As in the experiments with shake cultures the cell strains differed greatly from each other with respect to hydroxylation capacity. Despite some deviations, in general the results obtained with shake cultures could be reproduced in the bioreactor.

In contrast to the experiments with shake cultures, strain 72L produced only low amounts of β-methyldigoxin, if β-methyldigitoxin was added during the growth phase and pH was not regulated. Instead, the main product was purpurea glycoside A. If, however, pH was adjusted at 6.0, then only hydroxylation, occurred, and after 24 days of incubation about 20 mg l^{-1} β-methyldigoxin was formed. In experiments with strain E1, pH had to be raised even to 7.0, but because of cell lysis β-methyldigoxin formation was very slight[55,75–78]. In experiments with strains 347, 291 and 287 regulation of pH during growth phase (or stationary phase) was not necessary for β-methyldigoxin production. Whereas with strain 347 β-methyldigoxin formation was only low[79], in experiments with strain 291, however, 1.96 g of crystallized β-methyldigoxin could be isolated after 11 d of incubation with β-methyldigitoxin, which was added five times in doses of 800 mg during incubation[76]. Higher biotransformation rates were achieved with strain 287. This strain can transform repeated incubations of 40 mg l^{-1} of medium over a period of two days. When starting on day 5 of cultivation β-methyldigitoxin dissolved in ethanol was added every second day, 400 mg l^{-1} β-methyldigoxin were produced after 26 d of incubation. The same product yield was found using a constant substrate level of 40 mg l^{-1} during the whole day, but already after 15 d. A further enhancement was possible by exchanging ethanol by methanol as solvent for β-methyldigitoxin, and formation of 475 mg l^{-1} of β-methyldigoxin was observed during 14 d of incubation (Fig. 46).

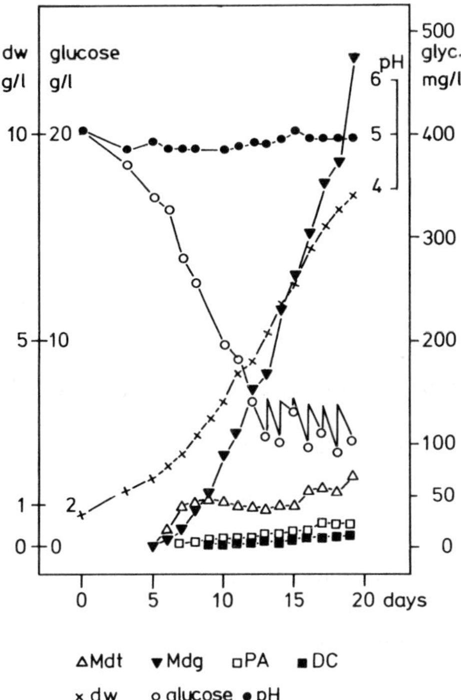

Fig. 46 Biotransformation of β-methyldigitoxin in a 20l-airlift reactor using a level of about 40 mg l^{-1} of β-methyldigitoxin in the medium. From day 14 glucose is fed to a level between 5 and 7.5 g l^{-1} β-Methyldigitoxin dissolved in 80% methanol; 24 °C; 350 Nl air h^{-1} (Helmbold[79])

Besides β-methyldigoxin, low amounts of PA and DC are formed too. Whereas both these compounds are found in the tissue fraction, about 90% of β-methyldigoxin can be extracted from the medium. Substrate and products — if at all — are degraded by strain 287 only at a negligible rate. Higher substrate levels than 40 mg l^{-1} turned out to be toxic for the cells, lower levels were insufficient. Furthermore, product formation could not be enhanced by changing the amount of inoculum (0.4 or 1.6 g instead of 0.8 g l^{-1}), by raising aeration rate from 350 to 700 Nl air h^{-1} temperature from 24 to 30 °C or by starting incubation on day zero instead of day five.

A further strategy for improving this process — besides more detailed investigations about medium requirements — will be to transfer the results obtained into a 200 l pilot plant and to investigate the possibility of separating growth of the cells from the biotransformation process. The cells will be cultivated discontinuously for achieving a desired cell density. Part of the reactor volume will then be transferred to a smaller reactor, in which biotransformation will be performed. When in the first reactor the optimal cell density is reattained, a further reactor can be filled. Using this method, it will be possible to apply optimal conditions for each part of the process, as well for the growth of the cells as for the biotransformation reaction. The first basic experiments were successfull[80].

7 Conclusions

As can be seen from the examples given above plant cell cultures represent an important potential to perform biochemical reactions on organic compounds. Epoxidation has been reported (Sect. 4.1), ester formation as well as ester saponification (Sect. 2.2; 2.3; 4.1; 6), glycosylation (2.1; 5.2; 5.3; 5.4; 6; cf. Ref. 7), hydroxylation (2.1; 4.5; 5.1; 6), isomerization (5.1; 6.4.1.1), methylation and demethylation (2.; 3; 4.2; 4.3; 6.4.2), reduction of double bonds, carbonyl and aldehyde functions (4.4; 5.1; 6.1; 6.2; 6.3; 6.4.1.2) as well as oxidation (5.1; 6.1; 6.2; 6.3; 6.4.1.1).

This large biochemical potential of plant cell cultures for production of secondary compounds by biotransformation may be used by two different approaches. First, one can employ as substrates compounds normally not available to the plant, like synthetic compounds, analogs of intermediates or products from other species, and possibly be able to obtain products up to now unknown in nature and, eventually, such products with new chemical and biological properties. The formation of glycosides reported above is of interest in some cases for the production of new compounds in the field of cardenolides (Sect. 6.4.1.1; 6.4.1.2) as well as for the unnatural substrate biotransformations of Sect. 2.3 and 3. Similar investigations could become of much more importance.

Second, one can transform the natural intermediates of the plant and obtain a product also available from the plant grown in the field. If such a reaction should become of biotechnological interest, several prerequisites have to be fulfilled. Besides the importance of the special compound one should be sure that this will not be replaceable in near future by other products. The availability of the substrate has to be secured, and the special biotransformation reaction has to be impossible for microorganisms as well as the technical chemist. Furthermore according to calculations about profitability of secondary product formation by plant cell cultures[5] one should claim that there is a price difference of at least 500 US $ per kg between substrate and product. In the example of hydroxylation of β-methyldigitoxin to β-methyldigoxin these prerequisites are fulfilled, and also high-yield cell strains have been isolated. It has been shown in recent years that by screening of large numbers of cell strains and, eventually, by additional single cell cloning it is well possible to isolate such high-yield cell strains[2, 5, 55, 64, 81–83]. Simple analytical methods may enhance such successful isolation[2, 81]. After isolation of such high-yield cell strains in shake or even on agar culture the appropriate optimal conditions for cultivation and biotransformation (or secondary product formation) in the bioreactor have to be elucidated for each strain, because these conditions may differ from strain to strain[55, 78, 79].

Using this strategy, i.e. β-methyldigoxin formation could be increased up to a fourty fold. It is to be hoped that possible problems arising from transferring the laboratory results to large reactor volumes as well as those of strain instability, which is not reported for *Digitalis* but for other species[5, 83], can be solved. Practical applications of biotransformation of organic compounds by cell culture methods will then be possible.

8 Abbreviations

C.	*Catharanthus*	*L.*	*Lycopersicum*
D.	*Digitalis*	LC	lanatoside C
2.4-D	2.4-dichlorophenoxy acetic acid	Me	methyl
DC	deacetyllanatoside C	Mdg	β-methyldigoxin
dw	dry weight	Mdt	β-methyldigitoxin
dtx	digitoxose	NAA	naphthalene acetic acid
glu	glucose	PA	purpurea glycoside A
glyc.	glycosides	PB	purpurea glycoside B
		stram.	*stramonium*

9 Acknowledgements

The authors express their thanks to Prof. T. Furuya (Tokyo) for providing valuable information and to Miss. I. Schuller for help in preparation of the English version of the manuscript. The receipt of a research grant from the Federal Minister for Research and Technology (Bonn) sustaining experiments in biotransformation of cardiac glycosides in this laboratory is gratefully acknowledged.

10 References

1. Tabata, M.: *In*: Plant Tissue Culture and Its Bio-Technological Application. Barz, W., Reinhard, E., Zenk, M. H. (Eds.), p. 3. Berlin: Springer 1977
2. Zenk, M. H., El-Shagi, H., Arens, H., Stöckigt, J., Weiler, E. W., Deus, B.: *In*: Plant Tissue Culture and Its Bio-Technological Application. Barz, W., Reinhard, E., Zenk, M. H. (Eds.), p. 27. Berlin: Springer 1977
3. Alfermann, A. W., Reinhard, E.: *In*: Production of Natural Compounds by Cell Culture Methods. Alfermann, A. W., Reinhard, E. (Eds.), p. 3. München, Gesellschaft für Strahlen- und Umweltforschung 1978
4. Tabata, M., Ogino, T., Yoshioka, K., Yoshikawa, N., Hiraoka, N.: *In*: Frontiers of Plant Tissue Culture 1978. Thorpe, T. A. (Ed.), p. 213. The International Association for Plant Tissue Culture. University of Calgary, Calgary (Canada) 1978
5. Zenk, M. H.: *In*: Frontiers of Plant Tissue Culture 1978. Thorpe, T. A. (Ed.), p. 1. The International Association for Plant Tissue Culture. University of Calgary, Calgary (Canada). 1978
6. Ellis, B.: Lloydia *37*, 168 (1974)
7. Barz, W.: *In*: Plant Tissue Culture and Its Bio-Technological Application. Barz, W., Reinhard, E., Zenk, M. H. (Eds.), p. 153. Berlin: Springer 1977

8. Pilgrim, H.: Pharmazie *25*, 568 (1970)
9. Tabata, M., Ikeda, F., Hiraoka, N., Konoshima, M.: Phytochem. *15*, 1225 (1976)
10. Steck, W., Constabel, F.: Lloydia *37*, 185 (1974)
11. Carew, D. P., Bainbridge, T.: Lloydia *39*, 147 (1976)
12. Romeike, A.: Biochem. Physiol. Pflanzen *87*, 168 (1975)
13. Hiraoka, N. Tabata, M., Konoshima, M.: Phytochem. *12*, 795 (1973)
14. Elze, H., Teuscher, E.: In: Biochemie und Physiologie der Alkaloide. Abhdl. Dtsch. Akad. Wiss. (Berlin), 1971. Mothes, K., Schreiber, K., Schütte, H. R. (Eds.), p. 239. Berlin, Akademie Verlag 1972
15. Stohs, S. J.: J. Pharm. Sci. *58*, 703 (1969)
16. Indra, A., Staba, E. J.: Phytochem. *7*, 79 (1968)
17. Steck, W., Gamborg, O. L., Bailey, B. K.: Lloydia *36*, 93 (1973)
18. Grützmann, K. D., Schröter, H.-B.: Abhdl. Deutsch. Akadem. Wiss. (Berlin), Kl. Chemie, *3*, 347 (1966)
19. Furuya, T., Nakano, M., Yoshikawa, T.: Phytochem. *17*, 891 (1978)
20. Fairbairn, J. W., Handa, S. S., Gürkan, E., Phillipson, J. W.: Phytochem. *17*, 261 (1978)
21. Böhm, H., Rink, E.: Biochem. Physiol. Pflanzen *168*, 69 (1975)
22. Boder, G. B., Gorman, M., Johnson, J. S., Simpson, P. J.: Lloydia *27*, 328 (1964)
23. Carew, D. P., Krueger, R. J.: Phytochem. *16*, 1461 (1977)
24. Veliky, I. A.: Phytochem. *11*, 1405 (1972)
25. Veliky, I. A., Barber, K. M.: Lloydia *38*, 125 (1975)
26. Hsu, P. Y., O'Connel, F. D.: Lloydia *36*, 435 (1973)
27. Butcher, D. N.: *In*: Applied and Fundamental Aspects of Plant Cell, Tissue, and Organ Culture. Reinert, J., Bajaj, Y. P. S. (Eds.), p. 668. Berlin: Springer 1977
28. Itokawa, H., Takeya, K., Akusa, M.: Chem. Pharm. Bull. *24*, 1681 (1976)
29. Itokawa, H., Takeya, K., Mihashi, S.: Chem. Pharm. Bull. *25*, 1941 (1977)
30. Takeya, K., Itokawa, H.: Chem. Pharm. Bull. *25*, 1947 (1977)
31. How, P., Macrae, A. R.: (unpublished), citated by Street, H. E.: In: Biosynthesis and Its Control in Plants. Milborrow, B. V. (Ed.), p. 121. London–New York, Academic Press 1973
32. Meehan, T. D., Coscia, C. J.: Biochem. Biophys. Res. Comm. *53*, 1043 (1973)
33. Suga, T., Hirata, T., Hirano, Y., Ito, T.: Chemistry Letters, p. 1245 (1976)
34. Aviv, D., Galun, E.: Planta Medica *33*, 70 (1978)
35. Furuya, T.: *In*: Frontiers of Plant Tissue Culture 1978. Thorpe, T. A. (Ed.), p. 191. The International Association for Plant Tissue Culture. University of Calgary, Calgary (Canada) 1978
36. Reinhard, E.: *In*: Tissue Culture and Plant Science 1974. Street, H. E. (Ed.), p. 433. London–New York, Academic Press 1974
37. Stohs, S. J.: *In*: Plant Tissue Culture and Its Bio-Technological Application. Barz, W., Reinhard, E., Zenk, M. H. (Eds.), p. 142. Berlin: Springer 1977
38. Gallili, G. E., Yagen, B., Mateles, R. I.: Phytochem. *17*, 578 (1977)
39. Yagen, B., Gallili, G. E., Mateles, R. I.: Appl. Environm. Microbiol. *36*, 213 (1978)
40. Hirotani, M., Furuya, T.: Phytochem. *14*, 2601 (1975)
41. Graves, J. M. H., Smith, W. K.: Nature *214*, 1248 (1967)
42. Furuya, T., Hirotani, M., Kawaguchi, K.: Phytochem. *10*, 1013 (1971)
43. Stohs, S. J., El-Olemy, M. M.: Lloydia *35*, 81 (1972)
44. Hirotani, M., Furuya, T.: Phytochem. *13*, 2135 (1974)
45. Stohs, S. J., El-Olemy, M. M.: J. Steroid. Biochem. *2*, 293 (1971)
46. Weber, N.: Z. Pflanzenphysiol. *87*, 355 (1978)
47. Farnsworth, N. R., Morris, R. W.: Amer. J. Pharm. *148*, 46 (1976)
48. Görlich, B.: Planta Medica *23*, 39 (1973)
49. Charney, W., Herzog, H. L.: Microbial Transformation of Steroids. New York–London, Academic Press 1967
50. Iizuka, H., Naito, A.: Microbial Biotransformation of Steroids and Alkaloids. University of Tokyo Press, Tokyo, and University Park Press, State College, Pennsylvania, 1969
51. Nozaki, Y., Mayama, M., Akaki, K., Satoh, D.: Agr. Biol. Chem. *29*, 783 (1965)
52. Güntert, Th. W., Linde, H. H. A.: Experientia *33*, 697 (1977)
53. Stohs, S. J., Staba, E. J.: J. Pharm. Sci. *54*, 56 (1964)

54. Stohs, S. J., Rosenberg, H.: Lloydia *38*, 181 (1976)
55. Alfermann, A. W., Boy, H. M., Döller, P. C., Hagedorn, W., Heins, M., Wahl, J., Reinhard, E.: *In*: Plant Tissue Culture and Its Bio-Technological Application. Barz, W., Reinhard, E., Zenk, M. H. (Eds.), p. 125. Berlin: Springer 1977
56. Döller, P. C.: Biotransformation von Cardenoliden. Vergleichende Untersuchungen mit Zellkulturen von *Thevetia neriifolia* und *Digitalis lanata*. Doctoral Thesis, Fachbereich Pharmazie, Universität Tübingen, 1978
57. Jones, A., Veliky, I. A., Ozubko, R. S.: Lloydia *41*, 476 (1978)
58. Veliky, I. A., Jones, A.: Poster presented at the 4th Int. Congr. Plant Tissue and Cell Culture. Book of Abstracts, abstract no. 700. Calgary (Canada), August 20–25, 1978
59. Döller, P. C., Alfermann, A. W., Reinhard, E.: Planta Medica *31*, 1 (1977)
60. Alfermann, A. W.: (Unpublished)
61. Furuya, T., Hirotani, M., Shinohara, T.: Chem. Pharm. Bull. *18*, 1080 (1970)
62. Furuya, T., Syono, K., Kojima, H., Hirotani, M., Ikuta, A., Hikichi, M., Kawaguchi, K., Matsumoto, K.: Proc. IV. IFS: Ferment. Technol. Today, p. 705, 1972
63. Reinhard, E., Boy, H. M., Kaiser, F.: Planta Medica Suppl., p. 163, 1975
64. Heins, M.: *In*: Production of Natural Compounds by Cell Culture Methods. Alfermann, A. W., Reinhard, E. (Eds.), p. 39. München, Gesellschaft für Strahlen- und Umweltforschung 1978
65. Tulecke, W., Nickel, L. G.: Science *130*, 863 (1959)
66. Mandels, M.: Adv. Biochem. Engin. *2*, 201 (1972)
67. Puhan, Z., Martin, S. M.: Progr. Industrial Microbiol. *9*, 14 (1971)
68. Kato, A., Shimizu, Y., Nagai, S.: J. Ferment. Technol. *53*, 744 (1975)
69. Kato, A., Kawazoe, S., Iijima, M., Shimizu, J.: J. Ferment. Technol. *54*, 82 (1976)
70. Yasuda, S., Satoh, K., Ishii, T.: Proc. IV. IFS: Ferment. Technol. Today, p. 697, 1972
71. Noguchi, M., Matsumoto, T., Hirata, Y., Yamamoto, K., Katsuyama, A., Kato, A., Azechi, S., Kato, K.: *In*: Plant Tissue Culture and Its Bio-Technological Application. Barz, W., Reinhard, E., Zenk, M. H. (Eds.), p. 85. Berlin: Springer 1977
72. Bligny, R.: Plant Physiol. *59*, 502 (1977)
73. Wagner, F., Vogelmann, H.: *In*: Plant Tissue Culture and Its Bio-Technological Application. Barz, W., Reinhard, E., Zenk, M. H. (Eds.), p. 245. Berlin: Springer 1977
74. Vcgelmann, H., Bischof, A., Pape, D., Wagner, F.: *In*: Production of Natural Compounds by Cell Culture Methods. Alfermann, A. W., Reinhard, E. (Eds.), p. 130. München, Gesellschaft für Strahlen- und Umweltforschung 1978
75. Wahl, J.: Fermentation von pflanzlichen Zellkulturen und 12β-Hydroxylierung und β-Methyldigitoxin durch Zellkulturen von *Digitalis lanata*. Doctoral Thesis, Fachbereich Pharmazie, Universität Tübingen 1977
76. Heins, M., Wahl, J., Lerch, A., Kaiser, F., Reinhard, E.: Planta Medica *33*, 57 (1978)
77. Heins, M.: Selektion 12β-hydroxylierender Zellstämme von *Digitalis lanata* und Umwandlung von β-Methyldigitoxin durch Fermenterkulturen. Doctoral Thesis, Fachbereich Pharmazie, Universität Tübingen 1978
78. Wahl, J.: *In*: Production of Natural Compounds by Cell Culture Methods. Alfermann, A. W., Reinhard, E. (Eds.), p. 48. München, Gesellschaft für Strahlen- und Umweltforschung 1978
79. Helmbold, U.: Versuche zur Steigerung der Hydroxylierung von β-Methyldigitoxin durch Fermenterkulturen von *Digitalis lanata* Zellstämmen. Doctoral Thesis, Fakultät für Chemie und Pharmazie, Universität Tübingen 1979
80. Reinhard, E., Heins, M., Helmbold, U., Wahl, J.: Poster presented at the 4th Int. Congr. Plant Tissue and Cell Culture. Book of Abstracts, abstract no. 702. Calgary (Canada), August 20–25, 1978
81. Ogino, T., Hiraoka, N., Tabata, M.: Phytochem. *17*, 1907 (1978)
82. Ohta, S., Kojima, Y., Yatazawa, M.: Agric. Biol. Chem. *42*, 1733 (1978)
83. Deus, B.: *In*: Production of Natural Compounds by Cell Culture Methods. Alfermann, A. W., Reinhard, E. (Eds.), p. 118. München, Gesellschaft für Strahlen- und Umweltforschung 1978

Metabolism of Steroids in Plant Tissue Cultures

S. J. Stohs
Department of Biomedicinal Chemistry, College of Pharmacy,
University of Nebraska Medical Center, Omaha, Nebraska 68105 USA

In the following article the current knowledge regarding the metabolism of steroids by cultured plant cells will be described. The various types of steroid biotransformation reactions which have been reported for plant tissue cultures will be reviewed. Furthermore, since many plant cell cultures are totipotent, that is, they possess all the information necessary for the replication and functioning of the whole plant, transformation reactions of potential industrial or scientific significance have also been described for some plants, further demonstrating the utility of plant tissue cultures. The results presented demonstrate that suspension cultures of plant cells can effect a wide-range of steroidal transformation reactions, several with industrial applicability.

1 Introduction

The utility of plant tissue cultures as a means of investigating biosynthesis and metabolism of steroids and other substances[1-5] has been well recognized since the rapid development of these techniques during the past 20 years. Biotransformation of particular substrates to more useful products by plant cells is now one of the most promising areas in the practical application of plant tissue cultures. Specific structural modifications of certain compounds may be effected more easily and economically than by chemical synthesis or microorganisms[4].

The microbial transformation of steroids has been well reviewed[6,7], the primary development of which occurred in the early 1950's. The transformation of steroids into more useful products by plant cells holds great economic promise, and recent developments suggest that the evolution of the applications of tissue cultures to this area are on the threshold of success. The selection of plant cell tissues, the development of cultures from high yielding parent plants, the regulation of pH and nutrients in a bioreactor, and the use of selective reaction inhibitors can all lead to higher yields of desired products.

Plant cell cultures, like microorganisms, can function as a convenient source of enzymes, and as a practical matter, the plant tissue culture of choice provides the necessary enzymes and cofactors for the desired transformation. It is necessary to provide the organism with a medium which is both suitable for growth and known to provide adequate levels of cofactors and enzymes. Furthermore, a knowledge of the nature of the enzymatic reactions of interest does alert the investigator to determine optimum temperature and pH, both of which are known to be important factors in establishing optimal enzymatic reaction rates.

As will become evident from the experimental results reported in this review and as is similar to microbial enzymes, the enzymes of plant cells involved in steroid biotransformations are not highly specific. Similar reactions can be effected with a variety of steroidal substrates, and it is doubtful that the cell possesses a distinct enzyme for each new substrate.

The transformations of steroids, carried out by intact plant cells, are believed to occur within the cell and not in the medium surrounding the cell. This situation is analogous to the transformation of steroids by microorganisms[6,7]. Most steroidal substrates have only modest solubilities in water and in the aqueous media used for plant cell suspension cultures. The steroid being transformed must dissolve to some extent in the medium so that diffusion through cell wall and presumably the cell membrane can occur, thereby coming in contact with the transforming enzymes. The situation with the biotransformation of glycosides as digitoxin and β-methyldigitoxin may be more complex. Steroidal glycosides are generally believed to undergo deglycosylation prior to entering the cells. However, β-methyldigoxin, the 12β-hydroxylation product of β-methyldigitoxin, is found in significant quantities both in the cell mass and in the medium of *Digitalis lanata* suspension cultures[5,8]. These results indicate that steroid glycosides are able to permeate the cell wall and membrane, unless the metabolizing enzymes are associated between the wall and membrane, and the substrate need only transverse the cell wall. Further studies are needed to clarify the cellular uptake of glycosidic substrates as the cardiac glycosides.

2 Culture Techniques

2.1 Culture Media

Plant tissue cultures may be grown in liquid nutrient media or on media solidified with agar. Tissues grown on agar usually form a callus or mass of unorganized cells, while liquid suspension cultures consist of mixtures of cell aggregates and single cells. The callus culture technique is favored for initiating and maintaining cell lines. Liquid suspension techniques are preferred for biosynthesis and biotransformation studies because this procedure provides higher growth rates, greater ease of substrate additions, superior control of the bioreactor environment, and more convenient vessels for growth.

A number of different media have been used to investigate the production, biosynthesis and biotransformation of steroids, including the media of Murashige and Skoog[9], White[10], Heller[11], and Nitsch[12]. The compositions of these four media in mg per liter are given in Table 1. In addition, 2–3 % sucrose or glucose is usually added

Table 1. Media formulations

Component ($mg\ l^{-1}$)	Medium			
	Murashige and Skoog[9]	White's (modified)[10]	Heller's[11]	Nitsch's H[12]
Macroelements				
$CaCl_2 \cdot 2\ H_2O$	440		75	166
$Ca(NO_3)_2$		208.5		
$NaNO_3$			600	
Na_2SO_4		200		
$NaH_2PO_4 \cdot 2\ H_2O$		18.7	141	
KH_2PO_4	170			68
KNO_3	1900	80		950
KCl		65	750	
FeNa · EDTA	36.7	4.59		36.7
NH_4NO_3	1650			720
$MgSO_4 \cdot 7\ H_2O$	370	720	250	185
Microelements				
$CoCl_2 \cdot 6\ H_2O$	0.025			
$NaMoO_4 \cdot 2\ H_2O$	0.25			0.25
$CuSO_4 \cdot 5\ H_2O$	0.025	0.001	0.03	0.025
KI	0.83	0.75	0.01	
H_3BO_3	6.2	1.5	1	10
$MnSO_4 \cdot 4\ H_2O$	22.3	7	0.1	25
FeNa · EDTA			1.36	
$ZnSO_4 \cdot 7\ H_2O$	8.6	3	1	10
MoO_3		0.0001		
$AlCl_3$			0.03	
$NiCl_2 \cdot 6\ H_2O$			0.03	
Amino acids & vitamins				
i-Inositol	100			100
Folic Acid				0.5
Nicotinic Acid	0.5	0.5		5
Thiamin HCl	0.1	0.1		0.5
Pyridoxine HCl	0.5	0.1		0.5
Glycine	2	3		2
Biotin				0.05

as a carbohydrate source. Modifications of these four media are also used to provide for specific requirements of some tissues. The most commonly employed medium for steroid biotransformation studies has been that of Murashige and Skoog[9] or modifications of this culture medium which contain a more complex vitamin mixture[13–15].

Most investigations are currently using from 0.1–1.0 mg 2,4-dichlorophenoxyacetic acid (2,4-D) per l^{-1} of medium to maintain undifferentiated tissue growth (see for example 16–20). In addition, from 0.1–2.0 mg l^{-1} kinetin is frequently used in addition to the 2,4-D[8,18,20]. Indolacetic acid (1.0 mg l^{-1}) has been used as a growth regulator in place of 2,4-D in combination with kinetin[8]. Some habituated cultures have been developed which require no growth regulator for maintaining undifferentiated growth. Relatively few studies have examined the influence of growth regulator concentration on the production and biotransformation of steroids. Diosgenin production was shown to be greatest in suspension cultures of *Dioscorea deltoidea* containing 0.10 mg l^{-1} 2,4-D as compared to 2,4-D concentrations up to 5 mg l^{-1} [21]. Most rapid growth was also achieved at the 0.10 mg l^{-1} level of 2,4-D. As a consequence, investigations involving the biotransformations of androst-4-en-3,17-dione[16], progesterone[23], cholesterol[24], and sitosterol[25] by *D. deltoidea* utilized cultures grown in the presence of 0.10 mg l^{-1} 2,4-D.

The pH of nutrient media is generally believed to be a critical factor in the production of secondary products as steroids and alkaloids[4,5,15]. However, it has been much neglected in tissue culture studies particularly with regard to the transformation of steroids. The usual practice is to adjust the pH of the medium to some value within the range of 5.0–6.0 during preparation of the medium. Drifts in pH do occur during course of the culture growth[8,21], and little is known of the influence of the actual pH on the ability of various cultures to biotransform or produce steroids. Heins *et al.*[8] reported that neither the pH or the growth period seemed to influence the ability of *Digitalis lanata* cell cultures to hydroxylate β-methyldigitoxin. The pH's ranged from approximately 4.5–6.0. Similarly, no correlation was seen between the pH in this same range and the conversion of cholesterol into diosgenin[21]. The 5β hydroxylation of digitoxigenin yielding periplogenin has been reported to occur more rapidly at pH 6.3 as compared to pH 4.8 and non-pH controlled conditions[22]. It is evident that the influence of pH on each biotransformation reaction involving a specific plant cell culture must be determined.

2.2 Aeration

Relatively little information is currently available on the effects of oxygenation and aeration on the production and biotransformation of steroids by plant tissue cultures. The regulation of aeration is not considered to be a serious problem[15,26]. However, plants are obligate aerobes, and as such, controlled oxygenation must be provided through aeration and/or mechanical agitation. Plant cells in suspension culture are thin-walled and extremely sensitive to shear stress. Mechanical impellers utilized as the only means of agitation may give poor growth yields due to resultant physical damage to the tissues. The use of airlift bioreactors in which aeration constitutes the only means of agitation has been successfully employed in the biotransformation of β-methyldigitoxin by cultures of *Digitalis lanata*[8]. These investigators did not report their rate of aeration. In the production of diosgenin by *Dioscorea deltoidea*, a combination of slow mechanical agitation with an impeller and aeration with sterile air has been effective[27]. The use of magnetic stirrers for agitation and aeration has been employed in the growth of suspension cultures[28, 29], although the technique has not been widely adopted, due apparently to the extent of tissue damage.

Radwan and Mangold[30] have reported that aerated suspension cultures of *Brassica napus* and *Glycine soja* were characterized by the production of relatively large proportions of steroids, while poorly aerated cultures were rich in the steroid precursor squalene. Similarly, the above aerated cultures contained large amounts of unsaturated fatty acids, whereas the poorly aerated cultures were characterized by a high content of very long chain fatty acids[31, 32]. The above results are easily explained in view of the fact that both the cyclization of squalene to steroids and the desaturation of fatty acids in biological systems require molecular oxygen, and demonstrates the importance of aeration to steroid production and metabolism.

2.3 Substrate Addition

As previously stated, most steroidal substrates have modest solubilities in aqueous media. To insure saturation of the culture medium and minimize this rate-limiting effect, the steroids are frequently dissolved in a wate-miscible solvent which results in the formation of a fine precipitation of the steroid upon addition to the aqueous medium. Ethanol (40–95%) has been particularly useful for the addition of steroid substrates to bioreactors (see for example[8, 16, 19, 23]). Ethanol is self-sterilizing and does not require additional sterilization procedures prior to its introduction into the bioreactor. Most cultures tolerate the ethanol addition to a final concentration of 0.5–1%. Ethanol can also be used to add suspensions of steroids. The introduction of steroids into a bioreactor in micronized form has not been reported for plant cell cultures. This technique is common in association with the microbial transformations of steroids[6], and offers an alternative means of providing a fine suspension of a steroid substrate for plant tissue cultures.

2.4 Illumination

The effect of light on the production and biotransformation of steroids has not been systematically investigated. In general, plant suspension cultures are commonly grown in diffuse light, although various investigators have reported that growth rates were similar when grown in the dark or in the presence of light[15, 33]. In reviewing the literature, no consistent pattern of light utilization was observed. Some cultures have been grown in the dark during steroid biotransformation studies[20]. Ordinary room light and its attendant variations have been employed[13, 16, 21, 23]. Many references do not state the lighting conditions utilized (see for example[8, 17, 19, 22]).

No substantial differences were detected in the lipid contents, composition of lipid classes and fatty acid patterns of lipids of *B. napus* and *G. soja* suspension cultures growth both in the dark and under continuous illumination[31].

Illumination does not appear to be an absolute requirement for the steroid biotransformation and biosynthetic studies reported to date. However, light may be required for certain reactions by selected plant cell cultures, and therefore until more information is available, this parameter should be assessed for each transformation system in question.

3 Classes of Chemical Reactions

The types of transformations of steroids which are currently known to be possible will be delineated according to classes of chemical reactions. These classes of reactions will be exemplified by biotransformation reactions involving specific substrates or the identification of endogenous steroid products from plant cells and plant cell cultures. Reactions are known to occur in plant tissues for the following classes of chemical reactions:

1. Oxidation
   1.1 Hydroxylation
   1.2 Oxidation of alcohols to ketones
   1.3 Dehydrogenation
2. Reduction
   2.1 Reduction of ketones to alcohols
   2.2 Reduction of double bonds
3. Isomerization
4. Glycosylation
5. Esterification
6. Side-Chain Cleavage

3.1 Oxidations

The most important single category in the transformation of steroids by plant cells is oxidation. The major impetus for the recent and rapid development of the entire field stems from the fact that hydroxylations constitute the most economically significant group of reactions. As a point of comparison, the steroid biotransformations by microorganisms with major technological application are hydroxylation, dehydrogenation, and oxidative side-chain degradation[6, 7].

3.1.1 Hydroxylations

Enzymatic hydroxylations of steroids by plant cells are believed to involve the same or similar mechanisms that occur in microorganisms and mammals. The enzymes involved are called the monooxygenases or hydroxylases, and involve the following type reaction —

$$S_t\text{—}H + X\text{—}H_2 + O_2 \rightarrow S_t\text{—}OH + H_2O + X$$

where S_t—H represents the steroidal substrate and X—H_2 is the electron donor. The entering oxygen atom is drived from molecular oxygen and not from water or other oxygen-containing compound present in the growth medium. Furthermore, the stereochemistry of the carbon atom that undergoes hydroxylation is maintained. In essence, the newly introduced hydroxyl group will have the same stereochemical configuration as the hydrogen atom which it replaced.

In most monooxygenase reactions, the substance that ultimately furnishes electrons to reduce one atom of oxygen to water is thought to be NADPH or NADH[36]. Such reactions have not been extensively studied in plants or plant cell cultures, and constitute a broad area for future research. Similarly, although the hemoprotein cytochrome P-450 has been demonstrated to exist in plants[37–39], its involvement in in the hydroxylation of steroids and other xenobiotics by plant cell cultures is assumed. The highly carcinogenic polycyclic aromatic hydrocarbon, benzo{α}pyrene, whose structure is related to the steroids, has been shown to be metabolized by plant suspension cultures of parsley (*Petroselinum hortense*)[40], soybean (*Glycine max*)[40], and *Chemopodium rubrum*[41] to a complete mixture of phenols, quinones and dihydrodiols. These reactions have been shown to be cytochrome P-450 dependent in mammalian systems[42]. The mechanism(s) involved in the enzyme-catalyzed hydroxylation of steroids in plant cells as well as by microbial and mammaliam tissues has not been completely elucidated.

In Figure 1 is given the basic cyclopentanoperhydrophenanthrene ring system of steroids, with the numbering system commonly employed. In this review, reference will be made to the various positions of the molecule at which biotransformations have been reported to occur by plant cells.

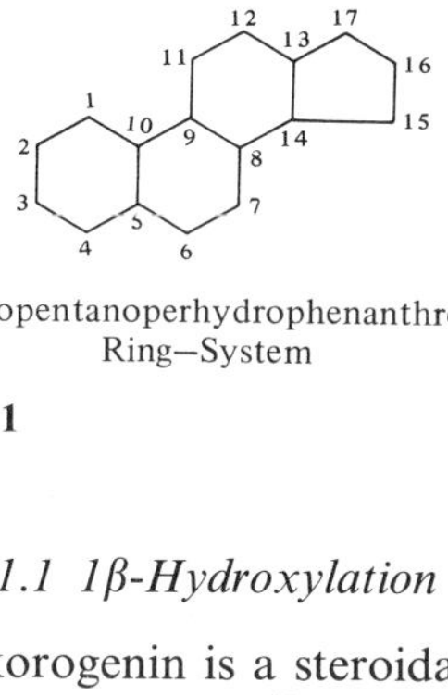

Cyclopentanoperhydrophenanthrene Ring–System

Fig. 1

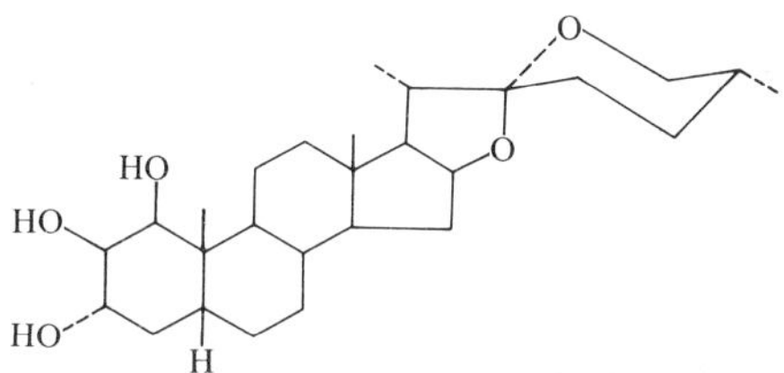

Tokorogenin

Fig. 2

3.1.1.1 1β-Hydroxylation

Tokorogenin is a steroidal sapogenin that has been isolated from tissue cultures of *Dioscosea tokoro*[43] (Fig. 2). It is a 1β,2β,3α-trihydroxy-steroid. Cultures of *D. tokoro* can accomplish the hydroxylations at positions C-1 and C-2 as demonstrated by the fact this culture can incorporate radiolabeled cycloartenol, cholesterol, cholest-5-ene-3β,16,26-triol and cholest-4-ene-3β,16,22,26-tetrol into tokorogenin[44]. Furthermore, hydroxylation and epimerization of diosgenin, a 3β-hydroxy steroidal sapogenin, to tokorogenin can be demonstrated with this plant cell culture[44].

Several other 1β-hydroxy steroids have been isolated from plants, further demonstrating the ability of plant cells to carry out this reaction. Such examples include ruscogenin from *Ruscus sp.*[45], cordylagenin and brisbagenin from *Cordyline cannifolia* and *C. stricta*[46], convallamarogenin from *Hosta sp.*[47] and reineckiagenin from *Reineckia carnea*[48].

The biosynthesis of digipurogenin I, a 1α-hydroxysteroid, from pregnenolone in intact Digitalis plants has been reviewed by Heftmann[49]. A 1α-hydroxylation has not been demonstrated to date in plant cell cultures.

3.1.1.2 2β-Hydroxylation

Yonogenin is a 2β,3α-dihydroxy sapogenin which can be produced by tissue cultures of *D. tokoro*[43]. It differs from tokorogenin only in that it lacks the 1β-hydroxyl group. Like tokorogenin, various steroids as cycloartenol, cholesterol and diosgenin are metabolized to yonogenin by this cell culture[44].

Other 2β-hydroxy steroidal sopogenins that have been isolated from plants include markogenin from *Yucca sp.*[50], samogenin from *Samuela carnerosana*[51], tokorogenin from *D. tokoro*[52], diotigenin[53] and isodiotigenin[43] from *D. tenuipes*.

A great variety of steroids with insect-molting hormone activity occur in plants. Ecdysone has been one of the most extensively investigated of these steroids (Fig. 3),

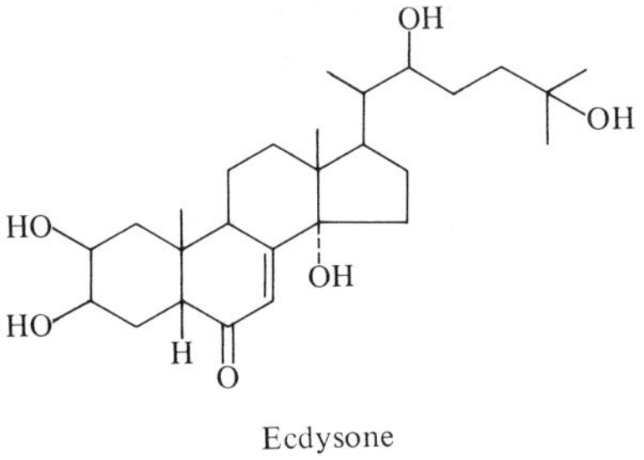

Ecdysone

Fig. 3a

Panasterone A

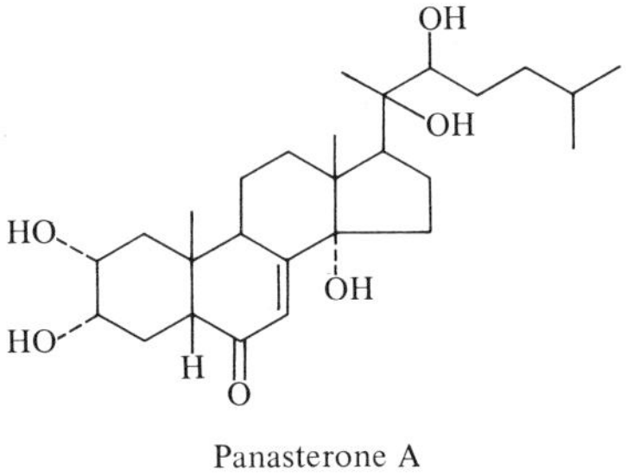

Fig. 3b

and the general group of insect-molting hormones is frequently referred to as the ecdysones. Most of the ecdysones identified to date contain a hydroxyl group at C-2. This hydroxyl group is usually in the β configuration as in ecdysone, but may be 2α as in the panasterone series of these compounds. Little information is available on the production of the ecdysones by plant cell cultures. However, the ability of many plant cells to accomplish the 2-hydroxylation is demonstrated by the wide distribution of the ecdysones. For example, of a total of 1056 plant species screened for insect-hormone molting activity, 54 were found to show this activity[54]. Some examples of plants which produce ecdysones include *Podocarpus macrophyllus*, *Trillium smallii*, *Helleborus niger*, *Taxus cuspidata*, *Lynchis miqueliana*, *Chrysanthemum morifolium*, *Ajuga decumbens*, and *Stachyurus praecox*.

3.1.1.3 2α-Hydroxylation

The ability of *Yucca glauca* tissue cultures to accomplish a 2α-hydroxylation was demonstrated by the isolation of gitogenin and manogenin which are 2α,3β-dihydroxy

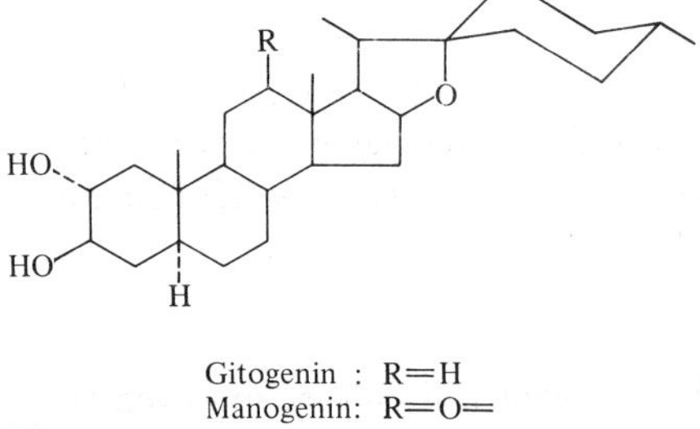

Gitogenin : R=H
Manogenin: R=O=

Fig. 4

spirostane derivatives[55] (Fig. 4). Maslinic acid is also a 2
while 3-epimaslinic acid is the 2α,3α-dihydroxy epimer. Both h
cell cultures of *Isodon japonicus*[56]. The ability of plants to effe
is further supported by the isolation of 2α-hydroxy sapogenin yu
and *Agave sp.*[57], and diotigenin from *Dioscorea tenuipes*[58], as wel
of the previously mentioned ecdysones.

3.1.1.4 3-Hydroxylation

More than 30 steroids have been isolated from plant cell cultures possess a 3-hydroxyl group[1]. No 3-hydroxylation of an exogenously provided substrate has been reported, and since all naturally occurring steroidal raw materials used commercially are oxygenated at the 3-position, the ability of plant cèlls to accomplish a hydroxylation at C-3 is of academic interest only.

3.1.1.5 4β-Hydroxylation

Isodiotigenin is the 4β-hydroxylated derivative of tokorogenin (Fig. 2). It has been isolated from seedlings of *D. tokoro*, but was not detected in tissue cultures of this plant[43]. Diotigenin has been isolated from plants of *D. tenuipes*[53], but not from cell cultures[43]. Other 4β-hydroxy sapogenins are kitigenin, which has been isolated from *Reineckia carnea* plants[58], and convallagenin B, which is produced by *Convallaria sp.*[59].

3.1.1.6 5β-Hydroxylation

Suspension cultures of *Daucus carota* will transform the cardenolide digitoxigenin to the single product periplogenin, which is 5β-hydroxydigitoxigenin[22] (Fig. 5). The most rapid rate of transformation occurred at pH 6.3. Digitoxigenin was inhibitory to this culture above a concentration of 50 mg l^{-1}.

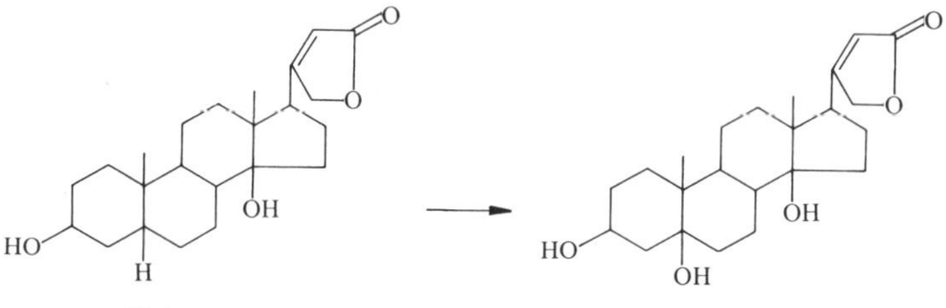

Digitoxigenin

Fig. 5a

Periplogenin

Fig. 5b

Other 5β-hydroxy cardenolides that are biosynthesized by plants include stophantidin and strophanthidol from *Strophanthus kombe*'[60], helleborin from *Helleborus atrorubens*[61], and cymarin and strophanthidin from *Apocynum cannabinum*[62] Tissue cultures of *A. cannabinum* have been shown to produce cardenolides similar but not identical to cymarin and strophanthidin[63].

3.1.1.7 6-Hydroxylation

The first demonstration of a 6β-hydroxylation by a plant cell culture involved the transformation of progesterone (pregn-4-en-3,20-dione) by *Lycopersicon esculentum*[19].

products actually isolated were pregn-4-ene-6β,11α-diol-3,20-dione (6β,11α-dihydroxyprogesterone) and pregn-4-ene-6β,11α-diol-3,20-dione (6β,14α-dihydroxy-progesterone) (Fig. 6). Progesterone was added to a final concentration of 300 mg l^{-1} of broth.

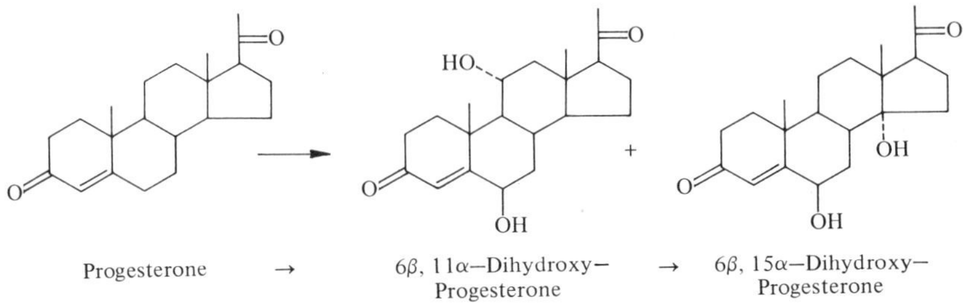

Progesterone → 6β, 11α–Dihydroxy–Progesterone → 6β, 15α–Dihydroxy–Progesterone

Fig. 6

Panaxatriol is the 6α-hydroxylated derivative of panaxadiol. Both sapogenins have been isolated from callus cell cultures of *Panax ginseng*[64)] (Fig. 7).

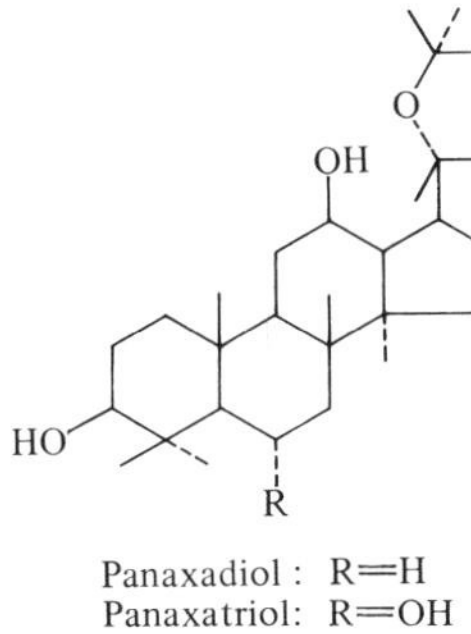

Panaxadiol : R=H
Panaxatriol: R=OH

Fig. 7

The ability to produce a C-6 oxidation in the form of a carbonyl has also been demonstrated for plant tissue cultures. Ergost-4-ene-3,6-dione, stigma-4-ene-3,6-dione, and stigma-4,22-diene-3,6-dione as well as the three corresponding steroids devoid of the 6-keto group have been isolated from cultures of *Glycine max* (soya)[65)] and *Stephania cepharantha*[66)]. Teng and Smith have shown that the enzyme soybean lipoxygenase will oxidize cholest-4-en-3-one to a mixture of epimeric cholest-4-en-6-hydroperoxide-3-ones. The 6-hydroperoxides were readily interconverted and decomposed to the corresponding epimeric 6-alcohol and 6-ketone derivatives.

Stigmast-4-en-6β-ol-3-one and campest-4-en-6β-ol-3-one have been isolated from the bark of *Melia azedarach*[68)]. Neochlorogenin is a 3β,6α-dihydroxy steroidal sapogenin that has been isolated from *Solanum pamiculatum*[69)].

3.1.1.8 7-Hydroxylation

The formation of the 7α and 7β-hydroxylated derivatives of cholesterol has been reported using root cultures of *Evomymus europaea* and *Digitalis purpurea*[70)]. In addition, 7-ketocholesterol was also detected. No other examples of a 7-hydroxylation of steroids by cultured plant tissues have been reported.

3.1.1.9 11α-Hydroxylation

The only report to date involving the formation of an 11α-hydroxylated steroid by a plant cell culture is the conversion of progesterone to 6β,11α-dihydroxyprogesterone by *Lycopersicon esculentum* as described above[19] (Fig. 6).

Tamusgenin is an 11-keto spirostane sapogenin, and both tamusgenin and diosgenin, the corresponding steroid devoid of the 11-keto group, have been isolated from the leaves of *Tamus edulis*[71].

3.1.1.10 12β-Hydroxylation

One of the most promising industrial applications of plant cell cultures involves the 12β-hydroxylation of cardiac glycosides. Suspension cultures of *Digitalis lanata* readily converted β-methyldigitoxin to β-methyldigoxin[5,8,72] (Fig. 8). In an initial

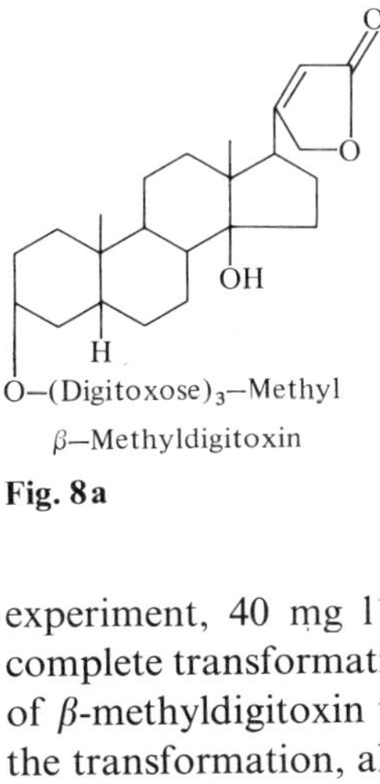

O–(Digitoxose)$_3$–Methyl

β–Methyldigitoxin

Fig. 8a

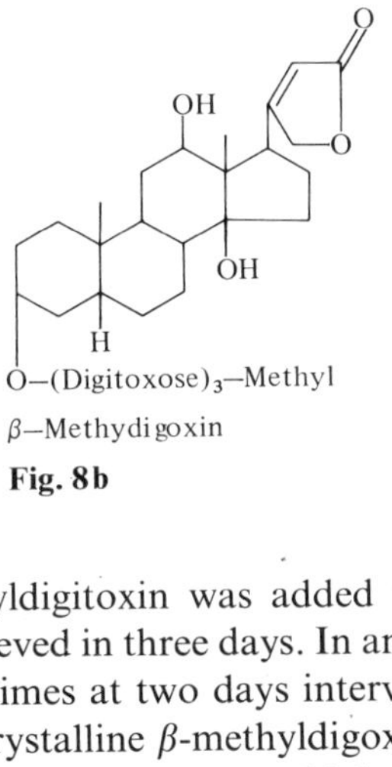

O–(Digitoxose)$_3$–Methyl

β–Methydigoxin

Fig. 8b

experiment, 40 mg l^{-1} of β-methyldigitoxin was added to a 20 l bioreactor with complete transformation being achieved in three days. In another experiment, 600 mg of β-methyldigitoxin was added 5 times at two days intervals. At the completion of the transformation, almost 2 g of crystalline β-methyldigoxin was recovered[8]. Strain selection of the cell culture was important to these high yields. These experiments demonstrate the commercial feasibility and potential of plant cell suspension cultures in the biotransformation of steroids.

Reinhard et al.[72] have also shown that digitoxin, acetyldigitoxin and lanatoside A all will undergo 12β-hydroxylation by cell suspension cultures of *D. lanata*. However, only the 12β-hydroxylation of β-methyldigitoxin occurred cleanly without other reactions as acetylation and glucosylation.

A number of 12β-hydroxy sapogenins have been isolated from plant cell cultures, demonstrating that other plants also possess the ability to accomplish a 12β-hydroxylation. Furuya et al.[64] have isolated the 12β-hydroxylated sapogenins panaxadiol and panaxatriol from callus tissue of *Panax ginseng* (Fig. 7). Tissue cultures of *Yucca glauca* produce the sapogenins gitogenin and manogenin[55] (Fig. 4). These two sapogenins differ only in that manogenin contains a 12-keto group and is therefore 12-ketogitogenin. The corresponding 12-hydroxy derivative was not detected in these cultures. However, a 12β-hydroxylated sapogenin, gloriogenin, has been isolated from the leaves of *Y. gloriosa*[73] and the seeds of *Y. glauca*[74]. Hecogenin is a 12-keto sapogenin which has been isolated from the leaves of several *Agave sp.*[51], and the leaves[75] and seeds[74] of *Y. glauca*. Willagenin is another 12-keto sapogenin, and it has been isolated from the leaves of *Y. filitera*[76].

3.1.1.11 14α-Hydroxylation

The 14α-hydroxylation of progesterone has been demonstrated by cell suspension cultures of *Lycopersicon esculentum* and *Vinca rosea*, giving rise to 14α-hydroxyprogesterone (preg-4-en-14α-ol-3,20-dione)[19]. Furthermore, 6β,14α-dihydroxyprogesterone was formed from progesterone by cell suspension cultures of *L. esculentum*[19] (Fig. 6).

As we have previously indicated, a wide variety of ecdysones occur in plants[54]. All of these compounds contain an α,β unsaturated carbonyl group and a 14α-hydroxyl moiety in addition to a 2-hydroxyl function[77] (Fig. 3). Although the concentrations of these steroids is much higher in plants than insects, they remained unrecognized until recently, in part because of their high water solubility. Their wide distribution[54] does demonstrate that many plant cells possess the ability to effect a 14α-hydroxylation.

3.1.1.12 15-Hydroxylations

Saikogenins A, B, and C are 15β-hydroxylated sapogenins while saikogenin D is a 15α-hydroxylated sapogenin. These sapogenins are produced by the plant *Bupleurum falcatum*. Uomori et al.[78] have reported that callus cultures of *B. falcatum* were unable to produce these four sapogenins, although the cultures were able to produce a number of other steroids including cholesterol, stigmasterol, campesterol, sitosterol, spinasterol and several Δ^7-steroids. Appropriate strain selection and environmental conditions may facilitate the production of the sapogenins by this culture and the hydroxylation at C-15.

Paravallarin is a steroidal alkaloid which has been isolated from *Paravallaris microphylla*, and possesses a 15α-hydroxy group[79].

3.1.1.13 16β-Hydroxylation

The 16β-hydroxylation of digitoxin to yield gitoxin has been reported for suspension cell cultures of *Digitalis purpurea*[80] (Fig. 9). The same hydroxylation by cultures of *D. lanata* was not observed[72].

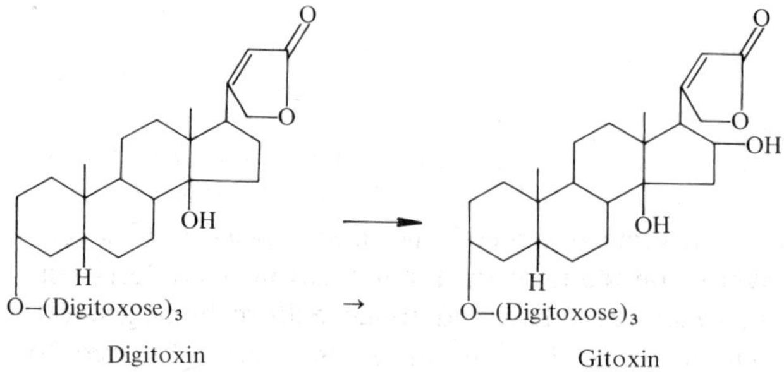

O
O
OH
H
O–(Digitoxose)$_3$
Digitoxin
→
O
O
OH
OH
H
O–(Digitoxose)$_3$
Gitoxin

Fig. 9

3.1.2 Oxidation of Alcohols to Ketones

The ability of plant cell suspension cultures to effect oxidation-reduction reactions involving the conversion of hydroxysteroids to ketosteroids has been widely demonstrated. Because a variety of steroidal substrates can be transformed by plant cell cultures, the evidence indicates that there is little substrate specificity among the cells producing hydroxysteroid dehydrogenases. The 3-hydroxyl group is very susceptible to oxidation. This reaction is depicted in Fig. 10, and in microbial systems, as for example by *Pseudomonas testosteroni*[81], has been characterized as being NAD-linked. NADP is believed to be involved in plant cells in this reaction.

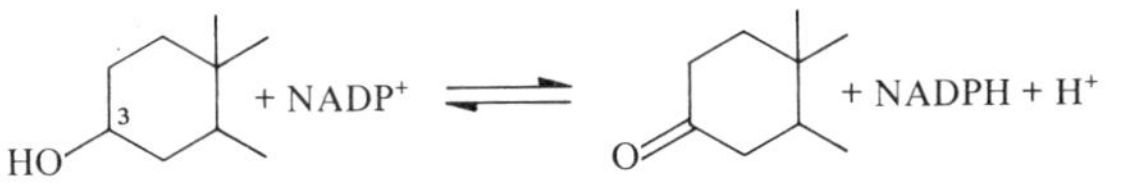

Fig. 10

Pregnenolone (pregn-5-en-3β-ol-20-one) can be readily transformed to progesterone (pregn-4-ene-3,20-dione) by cell cultures of *Digitalis purpurea*[82], *D. lutea*[82], *Nicotiana tabacum*[82, 34] and *Sophora angustifolia*[34] (Fig. 11). Leaf homogenates of *Cheiranthus cheiri*, *D. purpurea*, and *Strophanthus kombe'*[83] as well as intact leaves of *D. lanata*[84, 85] have also been reported to metabolize pregnenolone to progesterone. *D. purpurea* cell suspension cultures will readily interconvert 5β-pregnan-3β-ol-20-one and 5β-pregnane-3,20-dione[86].

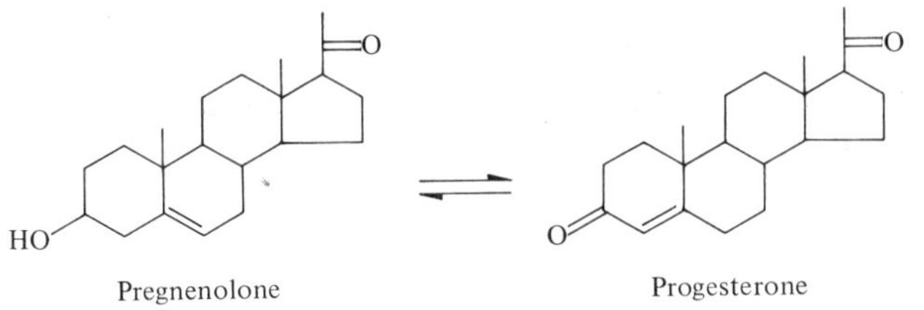

Fig. 11

Leaf homogenates of *Cheiranthus cheiri*, *Nerium oleander*, *Strophanthus kombe'*, *D. purpurea*, and *Corchorus capsularis* have been examined for their ability to metabolize 5α- and 5β-pregnan-3β-ol-20-one[87]. All tissues were able to oxidize the two substrates to the corresponding pregnane-3,20-dione. However, the 5α-pregnan-3β-ol-20-one was much more extensively metabolized, with the *N. oleander* and *C. capularis* being metabolically the most active.

Cholesterol (cholest-5-en-3β-ol) is rapidly converted to cholest-4-en-3-one by cell cultures of *Glycerine max* (soya) and *Brassica nupus* (rape)[20] (Fig. 12). This reaction is analogous to the pregnenolone to progesterone metabolic interconversion (Fig. 11). Molecular oxygen has been shown to be required for the formation of cholest-4-en-3-one from cholesterol[20]. *Cheiranthus cheiri* leaf and tissue culture homogenates, in the presence of an NADPH-generating system, have also been observed to metabolize cholesterol to cholest-4-en-3-one[88]. Leaf homogenates from *D. purpurea* were much less active in accomplishing this transformation, while homogenates of *N. oleander* leaves and *Dioscorea deltoidea* tissue cultures were devoid of metabolic activity under the same conditions[88].

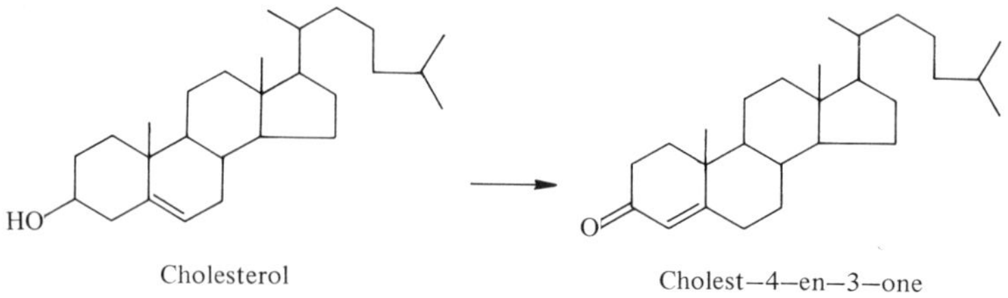

Cholesterol → Cholest–4–en–3–one

Fig. 12

In a biotransformation reaction analogous to that seen for cholesterol, sitosterol is transformed to sitost-4-en-3-one by *C. cheiri* leaf homogenates in the presence of an NADPH-generating system[89]. Under identical conditions, leaf homogenates of *Strophanthus kombe*' failed to metabolize sitosterol, while *D. purpurea* exhibited one-sixtieth of the metabolic activity of *C. cheiri*.

Hirotani and Furuya[35] have shown that cultured cells of *Nicotiana tabacum* are capable of transforming testosterone (androst-4-en-17β-ol-3-one) to androst-4-ene-3,17-dione (Fig. 13). The androst-4-ene-3,17-dione can then be further metabolized. The mechanism for this C-17 oxidation reaction is believed to be similar to the mechanism involved in the oxidation of the 3-hydroxy to the 3-keto (Fig. 10).

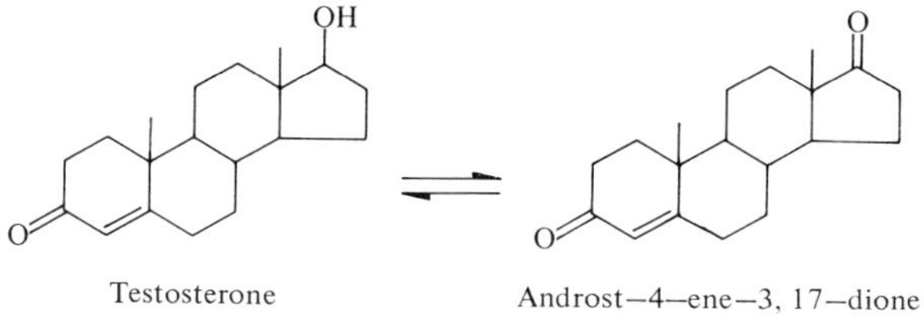

Testosterone ⇌ Androst–4–ene–3, 17–dione

Fig. 13

The reduction of digitoxigenin to the corresponding 3-keto drivative, digitoxigenone, has been observed for cell cultures of *D. lanata*, *D. purpurea*, and *Mentha spicata*[1,90].

3.1.3 Dehydrogenations

Few examples of desaturation of the A ring of steroids by plant cell enzymes exists. As shall be discussed under *Reductions*, the A ring, and the double bond at C-4 in particular, is readily reduced. These observations suggest that the oxidation to form the double bond does not readily occur, and thus the reaction is not easily reversible. Leaf homogenates of *Nerium oleander*, *Strophanthus kombe*', *Digitalis purpurea* and *Corchorus capsularis* in the presence of an NADPH-generating system and oxygen will transform 5α-pregnan-3β-ol-20-one via 5α-pregnane-3,20-dione into progesterone (pregn-4-ene-3,20-dione)[87]. The most active of these leaf preparations was *C. capsularis*.

3.2 Reduction

3.2.1 Reduction of Ketones to Alcohols

The reduction of ketones to hydroxyls is part of the equilibrium process described earlier. It is in essence the reverse of the oxidation of alcohols to carbonyl compounds (Fig. 10) and the same mechanism and principles are involved. Again one can assume that the enzyme(s) involved are very non-specific since a variety of substrates can be transformed, and the enzymes are found in many plant species.

The metabolism of 5α-pregnane-3,20-dione to 5α-pregnane-3β-ol-20-one was first demonstrated by Graves and Smith[82] (Fig. 14). This reduction of the 3-keto to the 3-hydroxy was effected by cultured cells of *Solanum tuberosum*, *Rosa sp.*, *Digitalis purpurea*, *D. lutea*, *Atropa belladonna*, *Nicotiana tabacum*, *N. rustica*, and *Hedera helix*[82]. This same transformation was reported by Furuya *et al.*[86] for *D. purpurea*, with the 5α-pregnan-3β-ol-20-one undergoing further metabolism. Similarly, *D. purpurea* suspension cultures will reduce 5β-pregnan-3,20-dione to 5β-pregnan-3β-ol-20-one[17], and the interconversion of 5β-pregnan-3β-ol-20-one and 5β-pregnan-3,20-dione has been demonstrated in *D. lanata* plants[91]. *Dioscorea deltoidea* cell cultures will convert progesterone (Fig. 11) to 5α-pregnan-3β-ol-20-one (Fig. 14)[23]. In this case, the 5α-pregnan-3,20-dione intermediate was not isolated.

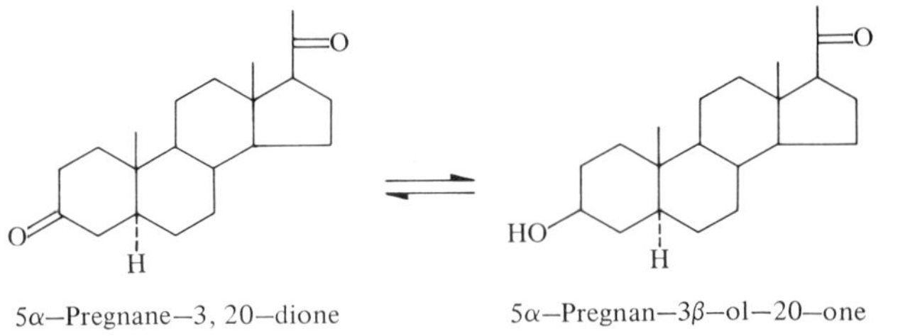

5α—Pregnane—3, 20—dione   5α—Pregnan—3β—ol—20—one

Fig. 14

Cholesterol (cholest-5-en-3β-ol) can be metabolized via cholest-4-en-3-one and 5α-cholestan-3-one to 5α-cholestan-3β-ol by suspension cultures of *Glycine max* and *Brassica napus*[20]. The reduction of cholest-4-en-3-one, derived from cholesterol, to 5α-cholestan-3β-ol by leaves of *Solanum tuberosum* has also been demonstrated[92].

The ability to reduce the 3-keto group of androst-4-en-3,17-dione was first demonstrated using suspension cultures of *Dioscosea deltoidea*, isolating 5α-androstan-3β-ol-17-one as one of the metabolic products which is formed via a 5α-androstan-3,17-dione intermediate[16]. Cell cultures of *Nicotiana tabacum* will reduce the 3-keto

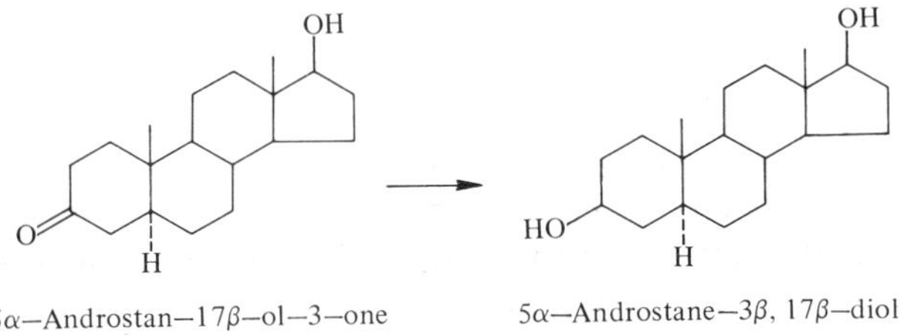

5α—Androstan—17β—ol—3—one   5α—Androstane—3β, 17β—diol

Fig. 15

group of 5α-androstane-3,17-dione and 5α-androstan-17β-ol-3-one to the 3β-hydroxyl, yielding 5α-androstan-3β-ol-17-one and 5α-androstane-3β,17β-diol, respectively[35] (Fig. 15). Furthermore, the 17-keto of androst-4-ene-3,17-dione can be reduced to give testosterone (androst-4-ene-3,17-dione) (Fig. 13) by suspension cultures of *N. tabacum*[35], while the 17-keto group of 5α-androstan-3β-ol-17-one can be reduced by cultures of *N. tabacum*[35] and *D. deltoidea*[16] to 5α-androstane-3β,17β-diol.

Graves and Smith[82] first demonstrated that plant tissue cultures were capable of reducing the 20-keto group of steroids to the 20-hydroxyl. Progesterone (pregn-4-ene-3,20-dione) (Fig. 11) was reduced to pregn-4-en-20α-ol-3-one[82] (Fig. 16). These

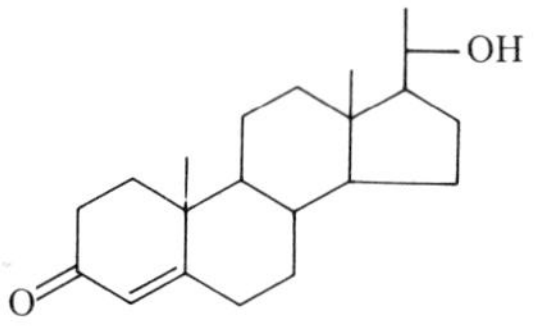

Pregn—4—en—20α—ol—3—one
Fig. 16a

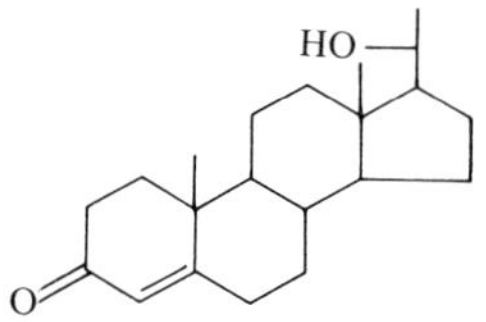

Pregn—4—en—20β—ol—3—one
Fig. 16b

results indicated that a stereospecific reduction at C-20 could be effected by proper culture selection. Suspension cultures of *D. purpurea* were also capable of transforming progesterone to both pregn-4-en-20α-ol-3-one and pregn-4-en-20β-ol-3-one, and also carried out the conversion of 5α-pregnan-3β-ol-3-one (Fig. 14) to both 5α-pregnane-3β,20α-diol and 5α-pregnane-3β,20β-diol[86] (Fig. 17).

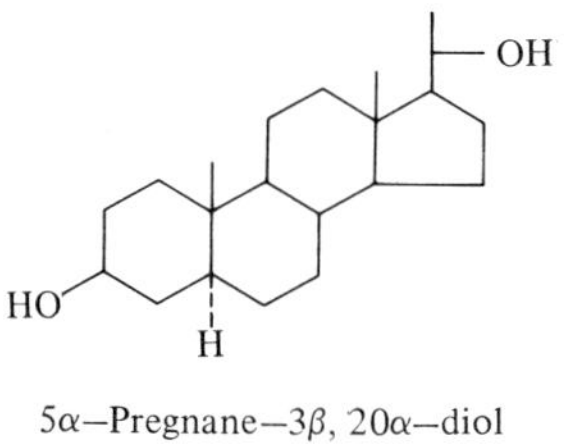

5α—Pregnane—3β, 20α—diol
Fig. 17a

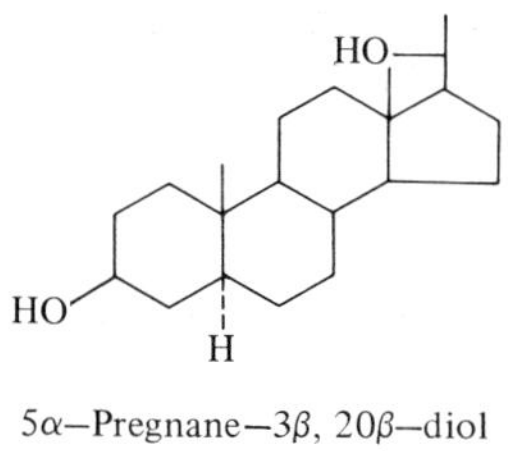

5α—Pregnane—3β, 20β—diol
Fig. 17b

Dioscorea deltoidea suspension cultures have been shown to be capable of metabolizing progesterone (Fig. 11) to 5α-pregnane-3β,20β-diol[23] (Fig. 17). The corresponding 20α isomer was not detected. When 5β-pregnan-3β-ol-20-one was used as the substrate for *D. purpurea* cell suspension cultures, 5β-pregnane-3α,20α-diol and 5β-pregnane-3β,20α-diol were isolated[17]. In addition, 5β-pregnane-3α,20α-diol was detected. The glucosides of these metabolites were also isolated. Furthermore, when normal cultures were compared with habituated cultures of *D. purpurea*, differences in the production of the two C-20 isomers could be demonstrated[17].

Leaf homogenates of *Cheiranthus cheiri*, *Nerium oleander*, *Strophanthus kombe*', *D. purpurea* and *Corchorus capsularis* have been examined for their ability to metabolize 5α-pregnan-3β-ol-20-one (Fig. 14) and 5β-pregnan-3β-ol-20-one[87]. All leaf homogenates were able to metabolize 5α-pregnan-3β-ol-20-one to 5α-pregnane-3β,20β-diol (Fig. 17). *C. cheiri* was most efficient in effecting this transformation. These leaf homogenates much less readily metabolized 5β-pregnan-3β-ol-20-one, with *D. purpurea* being most active in the transformation of this substrate to 5β-pregnane-3β,20β-diol. *C. cheiri* was not able to biotransform the 5β isomer[87].

3.2.2 Reduction of Double Bonds

Reduction of the double bond at C-4 of steroidal substrates has been demonstrated by a number of plant cell cultures (Fig. 18). The first demonstration that plant tissue cultures were able to effect this reaction involved the reduction of progesterone (pregn-4-ene-3,20-dione) (Fig. 11) to 5α-pregnane-3,20-dione (Fig. 14) by cultures of *Solanum*

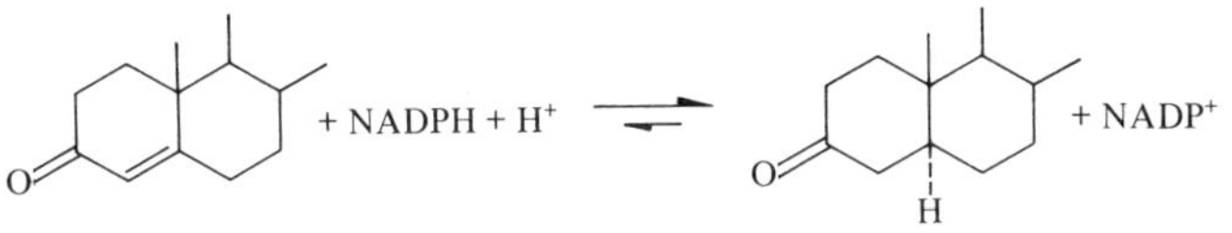

+ NADPH + H⁺ ⇌ + NADP⁺

Fig. 18

tuberosum, *Digitalis purpurea*, *D. lutea*, *Atropa belladona*, *Nicotiana tabacum*, *N. rustica* and *Hedera helix*[82]. Tissue cultures of *D. purpurea* were also shown by Furuya et al.[86] to metabolize progesterone to 5α-pregnane-3,20-dione.

Microsomes from cell cultures of *Cheiranthus cheiri* and *Dioscorea deltoidea* readily converted progesterone into 5α-pregnane-3,20-dione in the presence of an NADPH-generating system[93]. The reaction was time and enzyme dependent, with an optimum pH of 7. The NADPH-generating system could not be replaced by NADH, indicating that the reaction was NADPH dependent.

Leaf homogenates of *C. cheiri*, *Digitalis purpurea*, *Strophanthus kombe*' and *Corchorus olitorius* will metabolize progesterone to 5α-pregnane-3,20-dione in the presence of an NADPH-generating system[83]. Leaf homogenates of *C. cheiri* and *D. purpurea* will also metabolize pregnenolone (pregn-5-en-3β-ol-20-one) (Fig. 11) to progesterone and 5α-pregnane-3,20-dione. No 5β metabolites of either pregnenolone or progesterone could be detected.

Intact leaves of *D. lanata* will saturate the A ring of progesterone giving both 5α-pregnane-3,20-dione and 5β-pregnane-3,20-dione[94]. These same investigators also reported that leaves of *D. lanata* would metabolize pregn-5-en-3β-ol-20-one to 5β-pregnane-3,20-dione via progesterone[84].

Cell suspension cultures of *Glycine max* and *Brassica napus* will saturate the A ring of cholesterol (Fig. 12), giving rise to 5α-cholestan-3-one via cholest-4-en-3-one[20] (Fig. 19). The *Brassica* cultures were three to four times more active in facilitating this transformation than the *Glycine* cultures. NADPH was shown to be required when crude extracts of cell cultures were incubated with cholesterol, while NADH was not effective as a coenzyme for this reaction[20]. Homogenates of leaves of *D. purpurea*, *Cheiranthus cheiri* and *Strophanthus kombe*' will also metabolize cholest-4-en-3-one to 5α-cholestan-3-one in the presence of an NADPH-generating system[95]. Progesterone was capable of competing with cholest-4-en-3-one as the substrate, demonstrating a lack of specificity of the Δ^4-reductase enzyme system.

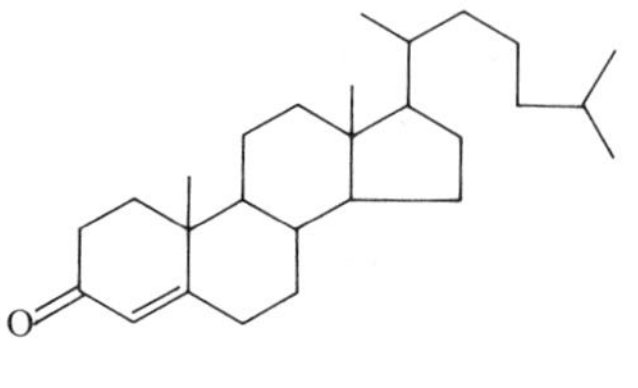

Cholest–4–en–3–one

Fig. 19a

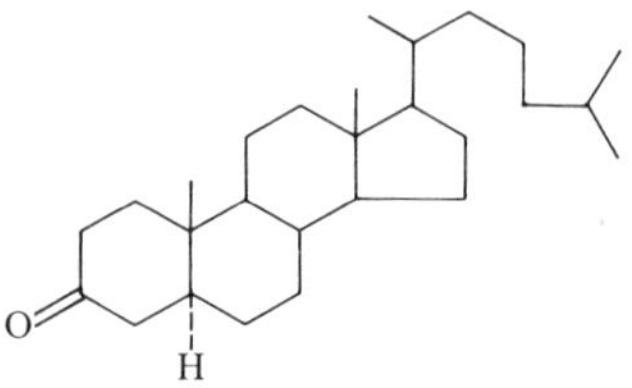

5α–Cholestan–3–one

Fig. 19b

Suspension cultures of *Dioscorea deltoidea* have been reported to reduce the A ring of androst-4-ene-3,17-dione[16]. The products isolated were 5α-androstan-3β-ol-17-one and 5α-androstane-3β,17β-diol (Fig. 15). Similarly, suspension cultures of *Nicotiana tabacum* transformed testosterone to 5α-androstan-3β-ol-17-one and 5α-androstane-3,17β-diol via androst-4-ene-3,17-dione (Fig. 13), which required reduction of the C-4 double bond[35].

3.3 Isomerization of Double Bonds

The primary isomerization reactions demonstrated to date for plant tissue cultures have involved the C-4 and C-5 positions, and most investigations have centered on the interconversion of pregnenolone (pregn-5-en-3β-ol-20-one) and progesterone (pregn-4-ene-3,20-dione) (Fig. 11). The conversion of pregnenolone to progesterone is favored. For example, *Digitalis purpurea*, *D. lutea*, and *Nicotiana tabacum* cell cultures were able to metabolize pregn-5-en-3β-ol-20-one to progesterone (pregn-4-ene-3,20-dione) while these cultures as well as cultures of *Atropa belladonna*, *Solanum tuberosum*, *N. rustica*, *Rosa sp.*, *Parthenocissus tricuspidata*, and *Hedera helix* were unable to carry out the reverse reaction[82]. Similarly, Furuya et al.[34] observed that suspension cultures of *N. tabacum* and *Sophora angustifolia* were able to convert pregnenolone to progesterone, but the conversion of pregesterone to pregnenolone was not detected. Earlier studies with intact leaves of *Digitalis lanata* also demonstrated that plant cells could metabolize pregnenolone to progesterone[84]. In addition, leaves of *D. lanata* have been shown to be able to biotransform progesterone to pregnenolone, although the yields were very low[94].

Cell suspension cultures of *Brassica napus* and *Glycine max* are capable of isomerizing the Δ^5 of cholesterol (cholest-5-en-3β-ol) to Δ^4 with cholest-4-en-3-one being the product isolated[20] (Fig. 12). We have shown that homogenates of *Cheiranthus cheiri* tissue cultures and leaves readily metabolized cholesterol to cholest-4-en-3-one[88], while homogenates of *Dioscorea deltoidea* tissue cultures and *Nerium oleander* leaves were unable to effect this transformation under identical conditions. Furthermore, *D. purpurea* leaf homogenates possessed only about one-seventieth the metabolic activity of the *C. cheiri* system. An NADPH generating system was required and greatest activity resided in the cytoplasmic fraction[88].

In an analogous reaction, leaf homogenates of *C. cheiri* have been shown to readily metabolize sitosterol (sitost-5-en-3β-ol) to sitost-4-en-3-one, utilizing NADPH[89]. Under identical conditions, homogenates of *Strophanthus kombe*' failed to transform sitosterol, while *D. purpurea* exhibited very low activity.

3.4 Glycosylation

Glycosylation reactions are known to readily occur in plant cells[96], and it is therefore not surprising to observe that plant cell cultures are capable of glucosylating steroids. Furthermore, many of the steroids isolated from plants and plant tissue cultures exist as conjugates and must be hydrolyzed before they can be extracted with common organic solvents[97].

The metabolic products of androst-4-ene-3,17-dione elaborated by *Dioscorea deltoidea* suspension cultures are retained in the tissue as conjugates, believed to be glucosides[16]. The biotransformation of testosterone by *N. tabacum* suspension cultures results in the production of a number of glucosylated products, including testosterone glucoside, 5α-androstan-3β-ol-17-one glucoside, and the 5α-androstane-3β,17β-diol-3- and 17-monoglucosides[35].

In the metabolism of progesterone by *D. deltoidea* suspension cultures, the hydroxylated products are also conjugated, existing as the glucosides in the tissue[23]. *Digitalis purpurea* cell cultures are capable of metabolizing 5β-pregnane-3,20-dione to the glucosides of 5β-pregnan-3β-ol-20-one and 5β-pregnan-3α-ol-20-one[17]. In addition to these products, this same culture was able to convert 5β-pregnan-3β-ol-20-one to 5β-pregnane-3β,20β-diol monoglucoside[17].

Several investigators have observed glucosylation of cardiac glycosides. Suspension cultures of *D. purpurea* were able to glucosylate digitoxin and gitoxin (Fig. 9) on the third digitoxose residue of the carbohydrate side-chain, yielding purpurea glycoside A and purpurea glycoside B, respectively[80]. Cell suspension cultures of *Digitalis lanata* also rapidly glucosylated digitoxin, forming purpurea glycoside A[72]. β-Methyldigitoxin, however, underwent selective 12β-hydroxylation in this system rather than glucosylation, indicating that substituents on the substrate molecule can direct the nature of the transformation reaction. Acetyldigitoxin was less efficiently glucosylated than digitoxin by this *D. lanata* culture[72].

3.5 Esterification

The ability of plant cell cultures to effect the esterification of steroids is not unexpected. Steroids and numerous other cell constituents exist as such in plants[96]. *Nicotiana tabacum* and *Sophora angustifolia* suspension cultures have been reported to biotransform progesterone to 5α-pregnan-3β-ol-20-one palmitate and pregnenolone to pregnenolone palmitate and 5α-pregnan-3β,ol-20-one palmitate[34]. Similarly, cultured cells of *N. tabacum* transformed testosterone to 5α-androstan-3β-ol-17-one palmitate and the dipalmitate of 5α-androstane-3β,17β-diol[35].

Cell suspension cultures of *Digitalis lanata* are able to both acetylate and de-acetylate digitalis glycosides added to the cultures[72]. Acetylation and de-acetylation of digitoxin occur rapidly. The acetylation of purpurea glycoside A also occurs, but less efficiently than the reaction with digitoxin. The deacetylation of acetydigitoxin and lanatoside A were shown to slowly take place by *D. lanata* cultures[72].

3.6 Side-Chain Cleavage

A majority of the worlds production of steroids involves partial synthesis from plant and animal sources[97]. In order to produce pharmaceutically useful products as corticosteroids, fertility control agents, sex hormones and anabolic compounds from such raw steroidal metarils as diosgenin, hecogenin, stigmasterol and cholesterol, side-chain cleavage must be effected. In the biosynthesis of cardiac glycosides in *Digitalis* and other plants, side-chain cleavage is required. A cell culture with the ability to rapidly cleave the side-chains of selected steroids would have an industrial and practical application. Unfortunately, the tissue cultures derived to date from *Digitalis purpurea*[27, 98], *Cheiranthus cheiri*[88], *Strophanthus kombe*[27], *Corchorus olitorius*[27] and *Corchorus capsularis*[27] are without side-chain cleavage activity. Pilgrim[98] has shown that although seedlings of *D. purpurea* possess cholesterol side-chain cleavage activity (Fig. 20), tissue cultures derived from these seedlings failed to retain

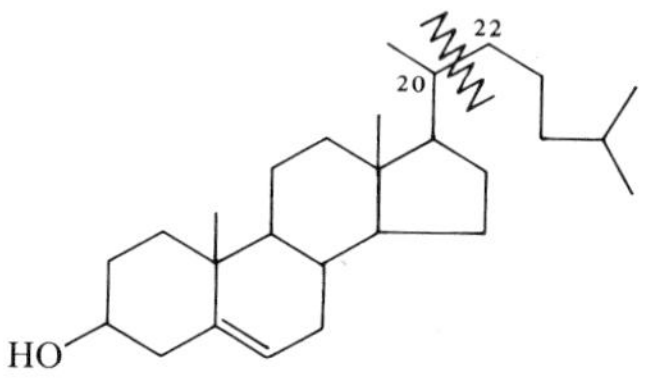

22
20
HO

Side—chain cleavage

Fig. 20

these enzymes. However, the potential exists, and possibly by appropriate strain selection, cell cultures with an active steroid side-chain cleavage enzyme system will be developed.

4 Conclusions

The most extensively investigated transformation reactions of steroids by plant cells involve hydroxylations. Such reactions currently offer great promise for the industrial application of cultured plant cells. Other steroid transformation reactions effected by plant cells include oxidation of hydroxyl groups to carbonyls, dehydrogenations, reduction of double bonds, reduction of carbonyls to hydroxyls, isomerizations, esterification and de-esterification reactions, and glucosylations. Research to date has shown that cultures which preferentially and efficiently accomplish a selected reaction can be developed. Culture and strain selection, not unlike that used for microorganisms, has measurably contributed to recent advances. Although a number of defined media have been developed on which a wide variety of cultures can be grown, little information is available on the effects of medium, pH, aeration, and temperature on the ability of specific cultures to carry out selected steroid biotransformations. Optimization of such conditions most certainly will enhance yields of desired products. The use of selective reaction inhibitors or modulators to block undesirable reactions and allow the accumulation of desired transformation products has been virtually unexplored.

As with microbial systems, the enzymes of plant cells involved in the biotransformation of steroids are not highly specific, and identical reactions can occur for a variety of related steroidal substrates by a given cell culture.

Plant tissue cultures have proven to be very useful for the study of steroid biosynthesis. However, only with recent advances has the industrial application of plant cell cultures become truly promising as a means of providing desired steroid transformation reactions.

5 Nomenclature

2,4-D 2,4-dichlorophenoxyacetic acid
NADP+ Nicotinamide adenine dinucleotide phosphate
NADPH Nicotinamide adenine dinucleotide phosphate, reduced
NADH Nicotinamide adenine dinucleotide, reduced

6 References

1. Stohs, S. J., Rosenberg, H.: Lloydia *38*, 181 (1975)
2. Stohs, S. J., *In*: Plant Tissue Culture and its Bio-Technological Application. Bartz, W., Reinhard, E., Zenk, M. H. (Eds), p. 142. Berlin, Springer 1977
3. Overton, K. W., Picken, D. J.: *In*: Progress in the Chemistry of Organic Natural Products. Herz, W., Gresebach, H., Kirby, G. W., (Eds), Vol. 34, p. 249. Berlin, Springer 1977
4. *In*: Plant Tissue Culture as a Source of Biochemicals. Staba, E. J., (Ed.). Cleveland, Chemical Rubber Company 1979
5. *In*: Production of Natural Compounds by Cell Culture Methods, Proc. Internat. Sympos. Plant Cell Culture: A. W. Alfermann, I. Reinhard (Eds.). München, Ges. Strahlen- u. Umweltforsch. 1978
6. Charney, W., Herzog, H. L.: Microbiol. Transformations of Steroids-A Handbook, New York, Academic Press 1967
7. Iizuki, H., Naito, A.: Microbial Transformations of Steroids and Alkaloids, Tokyo, University of Tokyo Press and University Park Press 1967
8. Heins, M., Wahl, J., Lerch, H., Kaiser, F., Reinhard, E.: Planta Med. *33*, 57 (1978)
9. Murashige, T., Skoog, F.: Physiol. Plant *15*, 473 (1962)
10. White, P. R.: The cultivation of Animal and Plant Cells. New York, The Ronald Press Co. 1954
11. Heller, R.: Ann. Sci. Nat. Bot. Biol. Veg. Paris *14*, 1 (1953)
12. Nitsch, J. P., Nitsch, C.: Science *163*, 85 (1969)
13. Kaul, B., Staba, E. J.: Lloydia *31*, 171 (1968)
14. Khanna, P., Staba, E. J.: Lloydia *31*, 180 (1968)
15. Staba, E. J.: Rec. Adv. Phytochem. *2*, 75 (1969)
16. Stohs, S. J., El-Olemy, M. M.: Lloydia *35*, 81 (1972)
17. Hirotani, M., Furuya, T.: Phytochem. *14*, 2601 (1975)
18. Tomita, Y., Uomori, A.: Phytochem. *13*, 729 (1974)
19. Yagen, B., Gallili, G. E., Mateles, R. I.: Appl. Environ. Microbiol. *36*, 213 (1978)
20. Weber, N.: Z. Pflanzenphysiol. Bd. *87*, 355 (1978)
21. Kaul, B., Stohs, S. J., Staba, E. J.: Lloydia *32*, 347 (1969)
22. Jones, A., Veliky, I. A., Ozubko, R. S.: Lloydia *41*, 476 (1978)
23. Stohs, S. J., El-Olemy, M. M.: Phytochem. *11*, 1397 (1972)
24. Stohs, S. J., Kaul, B., Staba, E. J.: Phytochem. *8*, 1679 (1969)
25. Stohs, S. J., Sabatka, J. J., Rosenberg, H.: Phytochem. *13*, 2145 (1974)
26. Mandels, M.: Adv. Biochem. Eng. *2*, 202 (1972)
27. Stohs, S. J., unpublished
28. Graebe, J. E., Novelli, G. D.: Exp. Cell Res. *41*, 509 (1966)
29. Wang, C. J., Staba, E. J.: J. Pharm. Sci. *52*, 1058 (1963)
30. Radwan, S. S., Mangold, H. K.: Adv. Lipid Res. *16*, 171 (1975)
31. Radwan, S. S., Mangold, H. K.: Chem. Phys. Lipids *14*, 87 (1975)

32. Radwan, S. S., Spener, F., Mangold, H. K., Staba, E. J.: Chem. Phys. Lipids *14*, 72 (1975)
33. Puhan, Z., Martin, S. M.: Prof. Ind. Microbiol. *9*, 13 (1971)
34. Furuya, T., Hirotani, M., Kawaguchi, K.: Phytochem. *10*, 1013 (1971)
35. Hirotani, M., Furuya, T.: Phytochem. *13*, 2135 (1974)
36. Coon, M. J., Strobel, H. W., Boyer, R. F.: Drug Metab. Disp. *1*, 92 (1973)
37. Young, O., Beevers, H.: Phytochem. *15*, 379 (1976)
38. Meehan, T. D., Coscia, C. J.: Biochem. Biophys. Res. Commun. *53*, 1043 (1973)
39. Rich, P. R., Lamb, C. J.: Eur. J. Biochem. *72*, 353 (1977)
40. Trenck, K. T. v. d., Sandermann, H.: Planta *141*, 245 (1978)
41. Harms, H., Dehnen, W., Monch, W.: Z. Naturforsch. *32*, 321 (1977)
42. DePierre, J. W., Ernster, L.: Biochim. Biophys. Acta *473*, 149 (1978)
43. Tomita, Y., Uomori, A., Minato, H.: Phytochem. *9*, 111 (1970)
44. Tomita, Y., Uomori, A.: Chem. Commun. *1971*, 284
45. Panova, D., Nikoloff, S., Achtardjieff, C.: Planta Med. *26*, 90 (1974)
46. Jewers, K., Burbage, M. B.: Steroids *24*, 203 (1974)
47. Takeda, K., Okanishi, T., Minato, M., Shimaoka, A.: Tetrahedron *21*, 2089 (1965)
48. Takeda, K., Okanishi, T., Minato, M., Shimaoka, A.: Tetrahedron *19*, 759 (1963)
49. Heftmann, E.: Lipids *9*, 626 (1974)
50. Wall, M. E., Eddy, C. R., Serota, S., Mininger, R. F.: J. Amer. Chem. Soc. *75*, 4437 (1953)
51. Marker, R. E., Wagner, R. B., Ulshafer, P. R., Wittbecker, E. L., Goldsmith, D. P. J., Ruof, C. H.: J. Amer. Chem. Soc. *69*, 2167 (1947)
52. Akahori, A.: Ann. Rep. Shionogi Res. Lab. *11*, 93 (1961)
53. Akahori, A.: Phytochem. *4*, 97 (1965)
54. Imai, S., Toyosato, T., Sakai, M., Sato, Y., Fujioka, S., Murata, E., Goto, M.: Chem. Pharm. Bull. *17*, 335 (1969)
55. Stohs, S. J., Sabatka, J. J., Obrist, J. J., Rosenberg, H.: Lloydia *37*, 504 (1974)
56. Tomita, Y., Seo, S.: J. Chem. Soc. Chem. Commun. *1973*, 707
57. Wall, M. E., Krider, M. M., Krewson, C. F., Eddy, C. R., Willaman, J. J., Corell, D. S., Gentry, H. S.: J. Amer. Pharm. Assoc. *43*, 1 (1954)
58. Takeda, K.: Chem. Pharm. Bull. *9*, 631 (1961)
59. Kimura, M., Tohma, M., Yoshizawa, I.: Chem. Pharm. Bull. *15*, 1713 (1967)
60. Sauer, H. H., Bennett, R. D., Heftmann, E.: Phytochem. *7*, 1543 (1968)
61. Tschesche, R., Brassat, B.: Z. Naturforsch. *20b*, 707 (1965)
62. Golab, T., Trabert, C., Jager, H., Reichstein, T.: Helv. Chem. Acta *42*, 2418 (1959)
63. Lee, P. K., Carew, D. P., Rosazza, J.: Lloydia *35*, 150 (1972)
64. Furuya, T., Kojima, H., Syono, K., Ishii, T., Uotani, K., Nishio, M.: Chem. Pharm. Bull. *21*, 98 (1973)
65. Weber, N.: Phytochem *16*, 1849 (1977)
66. Itokawa, H., Akasu, M., Fujita, M.: Chem. Pharm. Bull. *21*, 1386 (1973)
67. Teng, J. I., Smith, L. L.: J. Steroid Biochem. *7*, 577 (1976)
68. Nair, M. G., Chang, F. C. Phytochem. *12*, 903 (1973)
69. Ripperger, H., Schreiber, K., Budzikiewicz, H.: Chem. Ber. *100*, 1741 (1967)
70. Elzie, H., Pilgrim, H., Teuscher, E.: Pharmazie *29*, 10 (1974)
71. Barreira, R. F., Gonzalez, A. G., Rocio, J. A. S., Lopez, E. S.: Phytochem. *9*, 1641 (1970)
72. Reinhard, E., Boy, M., Kaiser, F.: Planta Med. Suppl. 1975, 163
73. Gonzalez, A. G., Barreira, R. F., Gonzalez, R. H., Salazar, J. A., Lopez, E. S.: Quimica *68*, 309 (1972)
74. El-Olemy, M. M., Sabatka, J. J., Stohs, S. J.: Phytochem. *13*, 489 (1974)
75. Blunden, G., Hardman, R. Phytochem. *8*, 1523 (1969)
76. Kenney, H. D., Wall, M. E.: J. Org. Chem. *22*, 468 (1957)
77. Heftmann, E.: Lipids *6*, 128 (1971)
78. Uomori, A., Seo, S., Tomita, Y.: Syoyakugaku Zasshi *28*, 152 (1974)
79. Tschesche, R.: Roc. Roy. Soc. Lond. B. *180*, 187 (1972)
80. Furuya, T., Hirotani, M., Shinohara, T.: Chem. Pharm. Bull. *18*, 1080 (1970)
81. Talalay, P.: *In*: The Enzymes. P. D. Boyer, H. Lardy, K. Myrback (Eds), Vol. 7, p. 177. New York, Academic Press 1963
82. Graves, J. M. H., Smith, W. K.: Nature *214*, 1248 (1967)

83. Stohs, S. J., El-Olemy, M. M.: Phytochem. *11*, 2419 (1972)
84. Sauer, H. H., Bennett, R. D., Heftmann, E.: Phytochem. *6*, 1521 (1967)
85. Caspi, E., Lewis, D. O.: Science *214*, 519 (1967)
86. Furuya, T., Kawaguchi, K., Hirotani, M.: Phytochem. *12*, 1621 (1973
87. Stohs, S. J.: Phytochem. *14*, 2419 (1975)
88. Stohs, S. J., El-Olemy, M. M.: J. Steroid Biochem. *2*, 293 (1971)
89. Stohs, S. J., El-Olemy, M. M.: Phytochem. *10*, 2987 (1971)
90. Stohs, S. J., Staba, E. J.: J. Pharm. Sci. *54*, 56 (1965)
91. Tschesche, R., Hombach, R., Scholten, H., Peters, M.: Phytochem. *9*, 1505 (1970)
92. Johnson, D. F., Waters, J. A., Bennett, R. D.: Arch. Biochem. Biophys. *108*, 282 (1964)
93. Stohs, S. J.: Phytochem. *8*, 1215 (1969)
94. Bennett, R. D., Sauer, H. H., Heftmann, E.: Phytochem. *7*, 41 (1968)
95. Stohs, S. J. Haggerty, J. A.: Phytochem. *12*, 2869 (1973)
96. Hitchcock, C., Nichols, B. W., *In*: Plant Lipid Biochemistry. p. 51. New York, Academic Press 1971
97. Hardman, R.: Trop. Sci. *11*, 196 (1969)
98. Pilgrim, H.: Phytochem. *11*, 1725 (1972)

Biochemistry of Lipids in Plant Cell Cultures

S. S. Radwan* and H. K. Mangold
Federal Center for Lipid Research, Institute for Biochemistry and Technology,
H. P. Kaufmann-Institute, D-4400 Münster,
Federal Republic of Germany

The composition of lipid classes and the patterns of constituent fatty acids of lipids from heterotrophic, photomixotrophic and photoautotrophic cultures are described and compared. The role of cultural conditions in stimulating the production of certain lipid classes and fatty acids is discussed. A review is given of the knowledge available on the biosynthesis and degradation of triacylglycerols, of various phospholipids and galactolipids, as well as of fatty acids and sterols. The advantages of cell culture techniques in basic and applied lipid research are illustrated and some further applications are suggested.

1 Introduction

In recent years much work has been done on the lipids of plant cell cultures. Two review articles[1,2] have appeared on this subject. The results of such studies are not only of academic interest, but also have practical value. It may be expected that cell cultures will be used in the future for the industrial production of valuable lipid compounds. If and when this stage of development is reached, cell cultures will be superior to intact

* Permanent address: Department of Botany, Faculty of Science, Ain Shams University, Cairo, Egypt

plants in many respects. For example, it will be possible to harvest the products desired after only a few weeks and not after months or even years as is the case with intact plants. Further, the crops that are produced under controlled conditions will not be subject to fluctuations due to weather, diseases and pests.

Plant cells in culture contain the genes that control the synthesis of valuable compounds as in the cells of intact plants. Compounds produced by intact plants are often not formed in cell cultures solely due to the fact that the environmental conditions favoring their biosynthesis have not yet been defined. No doubt extensive research work will be devoted to this problem.

In the present review, we shall be concerned mainly with metabolic aspects of lipids in plant cell cultures. We shall demonstrate that the biosynthesis of specific lipids can, in fact, be dramatically influenced by changing the environmental conditions during growth. This, certainly, holds true also for the biosynthesis of other natural products in cell cultures.

2 Composition of Lipids in Plant Cell Cultures

In the majority of studies of the composition of lipids in plant cell cultures, heterotrophic cultures have been investigated. Recently, photomixotrophic[3, 4] and photoautotrophic[5] cultures have been studied as well. In cell cultures lipids occur in cell membranes and cell organelles as well as in "oleosomes"; they also represent essential constituents of some enzyme systems.

2.1 Heterotrophic Cultures

Callus cultures and suspension cultures of many plant species have been found to contain between 3 and 8% total lipids, calculated on a dry weight basis[1, 2]. It appears that variations in the lipid contents and in the composition of lipid classes among cell cultures are primarily due to differences in cultural conditions, and, only to some extent, due to variations in plant species. Environmental conditions which stimulate protein synthesis such as the presence of high levels of amino acids in the nutrient medium, result in a reduction of total lipid contents in the cells[6]. The proportions of different lipid classes are dependent upon the age of the culture, the aeration, and other factors.

2.1.1 Lipid Classes

The lipid classes in heterotrophic cell cultures are qualitatively similar in their composition to those of actively growing nongreen organs of intact plants, but there are some quantitative variations. Cultures in the exponential growth phase[7, 8] are rich in phospholipids as well as sterols and complex lipids containing sterols, such as steryl esters, steryl glycosides and acylated steryl glycosides. Cultures in the late stationary phase of growth[9, 10] usually contain only low concentrations of phospholipids but high proportions of sterols and complex lipids containing sterols which may account for more than one half of the total lipids.

The predominant phospholipids in plant cell cultures are diacylglycerophosphocholines followed by diacylglycerophosphoethanolamines. Diacylglycerophosphoserines, diacylglycerophosphoinositols and phosphatidic acids are present at low levels.

Unlike the lipids from seeds, those from cell cultures contain only small proportions of triacylglycerols, usually between 5 and 10% of total lipids[1].

The lipids in heterotrophic cell cultures resemble those in nongreen organs of higher plants in that they contain very low levels of glycolipids which are characteristic constituents of chloroplasts[1].

Environmental conditions prevailing during growth affect the relative abundance of different lipid classes in cell cultures and their constituent fatty acids in many ways. Thus, the efficiency of aeration influences the relative concentrations of sterols and complex lipids containing sterols, and of the isoprenoid hydrocarbons[11]. Illumination does not affect the relative concentrations of lipid classes in heterotrophic cultures[11], and temperature has only a slight effect[12].

Lipid precursors and lipids which are added to the nutrient media are absorbed and metabolized by the cells. Thus, the lipid makeup of the cells is altered. Acetic acid and higher fatty acids are absorbed by cell cultures of soybean and are incorporated into phospholipids and triacylglycerols[13]. Sterols are also readily absorbed and metabolized[14,15].

Readers interested in more details on the composition of lipid classes in heterotrophic cell cultures are referred to two reviews mentioned earlier[1,2].

2.1.2 Constituent Fatty Acids

The fatty acid patterns of total lipids in heterotrophic cell cultures are similar to those in nongreen plants and plant organs[6,10,16]. Palmitic acid is the major saturated acid, whereas stearic acid is present only in small proportions. Oleic, linoleic and linolenic acids are the major unsaturated fatty acids. In most cases, the levels of linoleic acid, the major constituent fatty acid of phospholipids, are much higher than the levels of linolenic acid, the predominant constituent fatty acid of the galactolipids. For more details on the patterns of constituent fatty acids in total lipids and in the individual lipid classes of heterotrophic cell cultures the reader is again referred to two reviews[1,2].

Environmental conditions prevailing during growth affect the relative proportions of some constituent fatty acids of lipids in cell cultures. Thus, the lipids in cultures kept at low temperature contain higher levels of linoleic and linolenic acids than those in cultures incubated at ambient temperature[12].

Some of the unusual fatty acids that occur in the lipids of intact plants are present at much lower levels in heterotrophic cultures of these plants, whereas some are not detected at all. Thus, it has been found that the total lipids of heterotrophic cultures usually contain variable proportions of "very long-chain fatty acids" having up to 26 carbon atoms[6]. Erucic acid, which is a major constituent fatty acid in lipids from seeds of rape (*Brassica napus*) and nasturtium (*Tropaeolum majus*) is present only in traces in lipids from cell cultures of both plants[12,17,18]. It is of interest that lipids from seeds of nasturtium lack linoleic and linolenic acids, whereas in lipids from callus cultures of this plant those two polyunsaturated fatty acids are major constituents[12]. Ricinoleic acid, a hydroxy fatty acid which occurs at high levels in lipids from seeds of *Ricinus sp.*, is absent in lipids from callus cultures of this plant[19]. Similarly, in lipids from callus cultures of *Artemisia absinthium*[9] and *Petroselinum crispum*[20], epoxy fatty acids and petroselinic acid, respectively, are present in traces, if at all. The octadecenoic acids in the lipids of *Petroselinum crispum* cell cultures are the $\Delta 9$- and

Δ11-isomers, *i.e.*, oleic and vaccenic acid[20]. The octadecadienoic and octadecatrienoic acids, however, are pure linoleic and linolenic acids[20].

Substantial proportions of cyclopropene fatty acids occur in lipids from cell cultures of *Malva sp.*[21]. Small amounts of cyclopentenyl fatty acids occur in the lipids from cell cultures [22] and leaves[23] of species belonging to the family *Flacourtiaceae* whose seed lipids are known to contain these unusual fatty acids as major constituents[24].

2.2 Photoautotrophic Cultures

Recently, it has become possible to establish photomixotrophic[3,4] and photoautotrophic[25] cultures of some plant species. Such cultures cover their carbon requirements either partially (photomixotrophic cultures) or exclusively (photoautotrophic cultures) from carbon dioxide through photosynthesis.

2.2.1 Lipid Classes

Photomixotrophic and photoautotrophic cultures contain the same lipid classes that are present in heterotrophic cultures. In addition, lipids from photosynthetically active cell cultures[3-5] contain appreciable proportions of the lipid classes that are characteristic of green plants and plant organs[26,27], *i.e.*, monogalactosyldiacylglycerols (MGDG), digalactosyldiacylglycerols (DGDG), sulfoquinovosyldiacylglycerols (SQDG) and diacylglycerophosphoglycerols (PGP). The structures of these compounds are shown in Fig. 1.

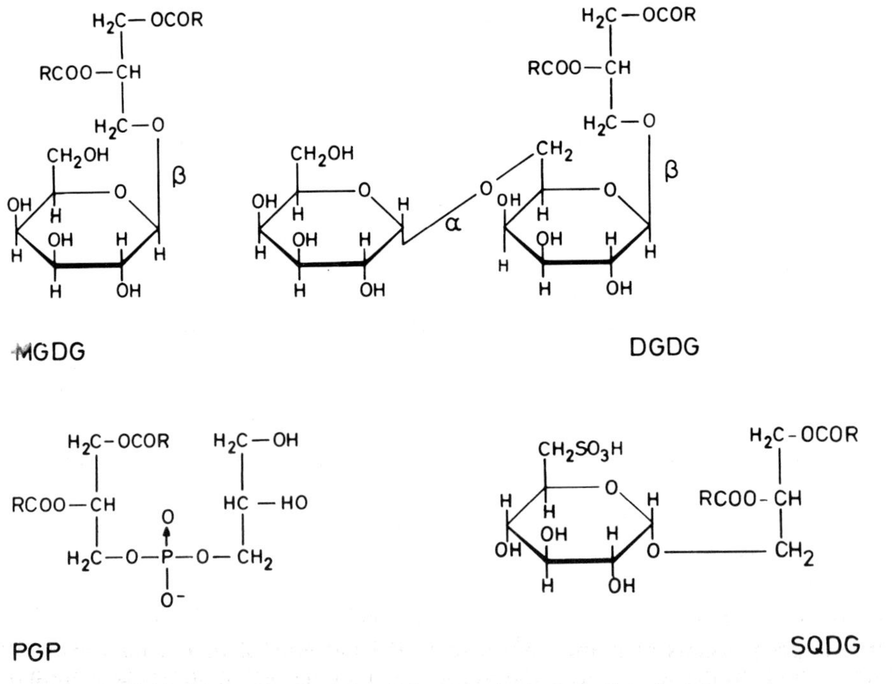

MGDG

DGDG

PGP

SQDG

Fig. 1 The lipid classes characteristic of photosynthetically active plant tissues
MGDG: Monogalactosyldiacylglycerols SQDG: Sulfoquinovosyldiacylglycerols
DGDG: Digalactosyldiacylglycerols PGP: Diacylglycerophosphoglycerols

In lipids from photoautotrophic cultures of *Chenopodium rubrum* the collective proportions of galactolipids amount to about 20 to 35% of the total lipids; sulfoquinovosyldiacylglycerols and diacylglycerophosphoglycerols constitute 1.5—5% and 3—6%, respectively[5]. Fig. 2 shows that photomixotrophic cultures, photoautotrophic cultures, and leaves of *Chenopodium rubrum* contain higher levels of galactolipids than heterotrophic cultures.

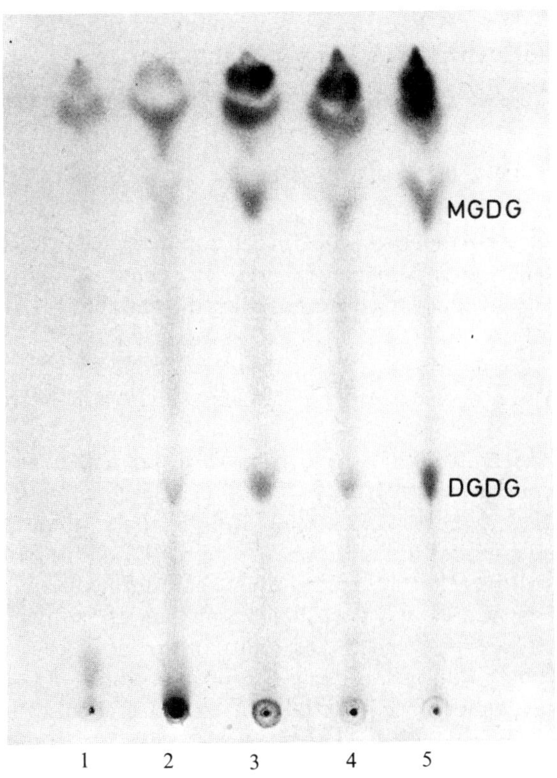

Fig. 2 A thin-layer chromatogram of galactolipids in cell cultures and leaves of *Chenopodium rubrum*

Sorbent: Silica Gel H impregnated with ammonium sulfate, 0.3 mm
Solvent: Acetone-benzene-water, 91:30:8 (by vol.)
Detection: α-Naphthol reagent
Galactolipids: MGDG: Monogalactosyldiacylglycerols, DGDG: Digalactosyldiacylglycerols
1: Heterotropic culture
2, 3: Photomixotrophic cultures
4: Photoautotrophic culture
5: Green leaves

2.2.2 Constituent Fatty Acids

Not only in the composition of lipid classes do photosynthetically active cell cultures resemble green organs of plants, but also in the patterns of their constituent fatty acids[3-5]. This similarity is expressed mainly in the relatively high levels of linolenic acid, which is characteristic of lipids in chloroplasts[27]. Photosynthetically active cell cultures and green plant organs contain linolenic acid predominantly esterified in monogalactosyldiacylglycerols and digalactosyldiacylglycerols[3–5, 26]. As mentioned before, the lipids from heterotrophic cell cultures usually contain much smaller proportions of this polyunsaturated fatty acid. All-*cis*-hexadecatrienoic and 3-*trans*-hexadecenoic acids occur in lipids from photosynthetically active cell cultures at much lower levels than in lipids from green leaves[3].

The onset of photosynthesis in plant cell cultures is accompanied by the stimulated synthesis of linolenic acid, as a constituent of monogalactosyldiacylglycerols and digalactosyldiacylglycerols, and by an enrichment in the levels of these galactolipids as well as of sulfoquinovosyldiacylglycerols and diacylglycerophosphoglycerols[28]. During the back shift from photoautotrophy to heterotrophy the proportions of these four lipid classes and their constituent linolenic acid decrease again[28].

3 Metabolism of Lipids in Plant Cell Cultures

The majority of studies published to date on the lipids of plant cell cultures have placed emphasis on the composition of lipid classes and the patterns of constituent fatty acids of lipids from these cultures. Publications that are concerned with the metabolism of lipids in cell cultures are still few in number.

3.1 Biosynthesis of Lipids

The biosynthetic activities of a plant cell culture change during growth and in response to a variety of factors. In the exponential phase of growth, the culture is active mainly in the biosynthesis of phospholipids[7, 8], whereas the same culture in the stationary phase accumulates mainly sterols and complex lipids containing sterols[9]. Photomixotrophic and photoautotrophic cell cultures that are grown under continuous illumination are much more active in the synthesis of galactolipids and their constituent linolenic acid than heterotrophic cultures[3–5, 28]. Such variability in the biosynthesis of lipids in cell cultures illustrates that the absence of certain natural products in the cells is, in many cases, only temporary, and is accounted for by the specific cultural conditions. In the following discussion we shall deal with several aspects of lipid metabolism, which confirm the validity of this statement.

3.1.1 Triacylglycerols

Plant cell cultures resemble actively growing organs of plants in that they synthesize only small amounts of triacylglycerols, usually between 5 and 10% of total lipids[1]. This holds true for both heterotrophic and photoautotrophic cultures, although unusually high levels of triacylglycerols, about 30% of total lipids, have been reported to occur in heterotrophic cultures of carrot (*Daucus carota*)[29]. Cell cultures enriched with triacylglycerols may be obtained by increasing the levels of certain growth regulators in the nutrient medium. Of special interest is that cell cultures of some plants contain "oleosomes", also called "spherosomes"[30], which are oil droplets imbedded in the cytoplasm of the cells. Electron micrographs of purified oleosomes are shown in Fig. 3.

Triacylglycerols from plant cell cultures may differ considerably in the patterns of their constituent fatty acids from those of seeds of the same plant species, in that they

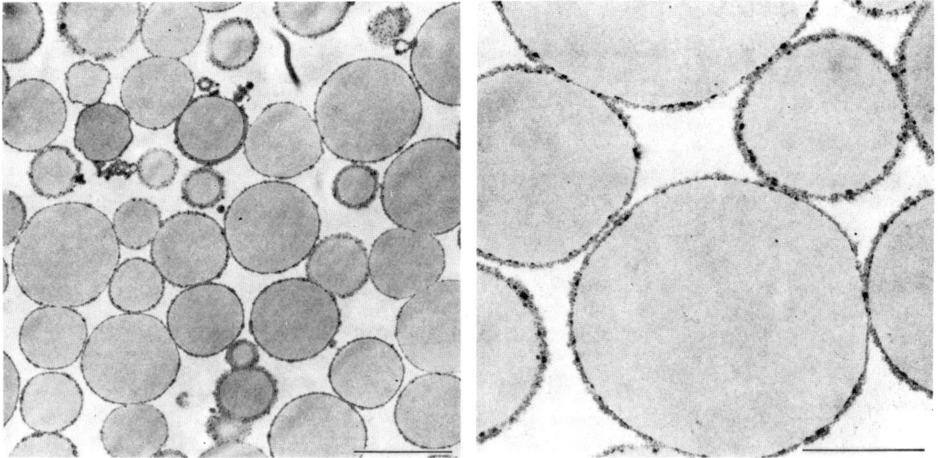

Fig. 3 Electron micrographs of purified oleosomes from cell cultures of *Daucus carota*. The bars indicate 1 μm (left) and 0.4 μm (right) [30]

contain higher levels of polyunsaturated fatty acids. As a rule, the unusual fatty acids characteristic of seed oils of some species are not synthesized and esterified in triacylglycerols or in any of the other lipid classes of cell cultures[16-20, 22].

It is of interest to note that under certain conditions the biosynthesis of triacylglycerols can be stimulated in heterotrophic cultures. Thus, tissue cultures of rape synthesize considerable proportions of triacylglycerols when they are treated with plant hormones that promote embryogenesis[31]. The pattern of fatty acids in the triacylglycerols becomes similar to that in the triacylglycerols from seeds, in as much as relatively high levels of unusual fatty acids appear. Of course, this does not imply that plant cell cultures can be used for the industrial production of oil. Such a process would certainly be very uneconomical, even if maximal biosynthesis of triacylglycerols was maintained[31]. These results are mentioned only to demonstrate that under proper conditions, plant cell cultures can synthesize large proportions of compounds that they normally produce in small amounts only.

The mechanism of biosynthesis of triacylglycerols in plant cell cultures apparently does not differ from that in cells of intact plants. Thus, the labelling pattern of fatty acids in soybean cell cultures growing in media that have been supplied with [1-^{14}C] acetate, is similar to that in intact plants[32]. [2-^{3}H]Glycerol is also incorporated into the triacylglycerols of plant cell cultures[28]. The results presented in Table 1 show that after only 16 minutes incubation of soybean cells with [1-^{14}C]acetate, most of the radioactivity appears in the oleic acid esterified in triacylglycerols. After one day, radioactivity becomes predominant in linolenic acid. These results suggest that linolenic acid is produced from oleic acid by desaturation *via* linoleic acid[33].

When [1-^{14}C]acetate is incorporated into triacylglycerols of cell cultures, the radioactivity appears in both the acyl moieties and the glycerol moiety[34]. After two days incubation of soybean cell cultures with radioactive precursors, the cells are found to have incorporated the radioactivity from ethanolamine, choline and serine not only in

Table 1. Incorporation of [1-^{14}C]acetate into fatty acids of triacylglycerols from soybean cell cultures[33]

Chain length: Number of double bonds	Duration of incubation with [1-^{14}C]acetate				
	16 (min)	45 (min)	85 (min)	155 (min)	22 (h)
16:0	26.4	23.1	18.8	15.5	15.1
18:1	60.5	55.7	48.2	36.5	6.4
18:2	10.0	17.6	26.1	33.9	18.8
18:3	3.1	3.6	6.9	14.1	59.7

Values are expressed as % of radioactivity in total fatty acids of triacylglycerols
5 day cells grown under continuous illumination were used

phospholipids, but also, surprisingly, in triacylglycerols[34]. Under these conditions, about one fourth to one third of the radioactivity in total lipids is incorporated into the acyl moieties of triacylglycerols. Obviously, ethanolamine, choline and serine are degraded to acetate which is then incorporated into fatty acids according to established mechanisms.

3.1.2 Phospholipids

A characteristic of plant cell cultures is that their phospholipid contents show considerable variations during growth and in response to changes in cultural conditions. Thus, cultures in the exponential growth phase contain higher levels of phospholipids than cultures in the stationary growth phase. For example, 48 h cell cultures of soybean, at the beginning of the exponential phase, contain about 5 times more diacylglycerophosphoethanolamines and diacylglycerophosphocholines than 192 h cultures, at the end of the exponential phase[8]. Cell cultures of sycamore (*Acer pseudoplatanus*), at the middle of the exponential phase, contain about 15 times more phospholipids than cells at the beginning of the stationary phase[7]. Also, 7 day cultures of soybean, at the end of the exponential phase, contain 6–7 times more phospholipids than 14 day cultures in the stationary phase[35]. The high levels of phospholipids in cells in the exponential growth phase is expected because during this phase new cells are continuously being produced. It is well known that phospholipids are main structural constituents of biological membranes and of different cell organelles.

The temperature prevailing during growth has some effect on the levels of phospholipids in plant cell cultures. At 5 °C the relative concentrations of phospholipids in cell cultures of rape are higher than at 30 °C[12].

The biosynthesis of phospholipids from radioactively labelled precursors has been investigated using soybean and carrot cell cultures. As a rule, the different precursors are incorporated rapidly into phospholipids and other lipid classes. When soybean cell cultures are incubated with [1-^{14}C]acetate, a major part of the radioactivity rapidly appears in the cell lipids; it remains fairly constant at a high level for 22 h[33]. Most of the radioactivity is incorporated into the cells' phospholipids. Also when other labelled precursors such as ethanolamine, choline, serine, inositol[28, 34] and glycerol[36] are incubated with cell cultures, they are incorporated rapidly into cell lipids, mainly into phospholipids.

Radioactivity from different precursors shows various rates of incorporation into the individual phospholipids. The amount of radioactivity which appears in individual phospholipids is dependent, not only on the identities of the phospholipids and precursors, but also on the duration of incubation of cells with the precursor, as well as on other factors. Thus, when 4 to 5 day cultures of soybean grown under continuous illumination are incubated with [1-^{14}C]acetate, radioactivity appears predominantly in diacylglycerophosphocholines and, to a lesser extent, in diacylglycerophosphoethanolamines of the cells during the first 2 h (Table 2). After 6 to 22 h, the radioactivity in these two classes of phospholipids decreases dramatically[33]. However, when 9 day cultures of soybean are incubated with [1-^{14}C]acetate, the major portion of the radioactivity is still in diacylglycerophosphocholines and in diacylglycerophosphoethanolamines, even after 48 h[34].

Table 2. Incorporation of [1-^{14}C]acetate into phospholipids of soybean cell cultures

Duration of incubation with [1-^{14}C] acetate	Phospholipids			Ref.
	Diacylglycero-phospho-ethanolamines	Diacylglycero-phospho-cholines	Other phospho-lipids	
16 min*	18.2	59.0	1.0	33)
45 min*	18.1	42.4	3.0	33)
85 min*	22.2	39.4	3.3	33)
155 min*	30.8	33.2	3.1	33)
6 h**	8.0	5.0	76.0	32)
22 h*	3.1	2.5	53.1	33)

Values are expressed as % of radioactivity in total lipids
* 5 day cells grown under continuous illumination were used
** 21 day cells grown under 8 h/day illumination were used

Similar results are obtained when soybean cells are incubated with [methyl-^{14}C]-choline, [2-^{14}C]ethanolamine or [3-^{14}C]serine as shown in Table 3.

Cell cultures of *Daucus carota* were reported to incorporate [methyl-^{3}H]choline and [2-^{3}H]inositol up to 100% in diacylglycerophosphocholines and diacylglycerophosphoinositols, respectively, whereas the radioactivity originating from [1-^{3}H]-ethanolamine was found to appear in both diacylglycerophosphoethanolamines and diacylglycerophosphocholines[29]. Working with cell cultures of soybean we have found that radioactivity from [2-^{14}C]ethanolamine, [methyl-^{14}C]choline and [3-^{14}C]serine is not incorporated into just a single class of phospholipids, but is distributed among several classes[34]. Irrespective of the radioactive precursor added, most of the radioactivity is incorporated into diacylglycerophosphocholines, and to a lesser extent in diacylglycerophosphoethanolamines. A smaller portion of radioactivity appears even in triacylglycerols, steryl esters, and galactolipids. When [3-^{14}C]serine is used as a precursor no detectable activity[34] (Table 3), or, according to another investigator[8], only 10% of activity in total lipids, appears in diacylglycerophosphoserines, whereas, diacylglycerophosphoethanolamines, diacylglycerophosphocholines, triacylglycerols,

Table 3. Incorporation of labelled precursors into phospholipids of plant cell cultures

Cell cultures	Incubation with the precursor [h]	Ethanolamine[a]		Choline[b]		[3-^{14}C] Serine		[2-^{3}H] Inositol	Ref.
		Diacylglycerophosphoethanolamines	Diacylglycerophosphocholines	Diacylglycerophosphoethanolamines	Diacylglycerophosphocholines	Diacylglycerophosphoethanolamines	Diacylglycerophosphocholines	Diacylglycerophosphoinositols	
*G. max**	48	19.0	46.2	6.4	55.3	20.9	48.7		[34]
*D. carota***	48	70.2	29.8		100			100	[29]

Values are expressed as % of radioactivity in total lipids

[a] [2-^{14}C] Ethanolamine was incubated with *G. max*, whereas [1-^{3}H] ethanolamine was incubated with *D. carota*

[b] [methyl-^{14}C] Choline was incubated with *G. max*, whereas [methyl-^{3}H] choline was incubated with *D. carota*

* 9 day cells grown in the dark were used

** 5 day cells grown in the dark were used

steryl esters and even the galactolipids are labelled[34]. A similar pattern of labelling is observed when [2-^{14}C]ethanolamine is used as a precursor[34].

It has been assumed that diacylglycerophosphoethanolamines and diacylglycerophosphocholines may be formed by decarboxylation of diacylglycerophosphoserines and subsequent methylation or by initial conversion of serine to ethanolamine and choline which are then incorporated into the respective phospholipids[8]. In this context, it is to be noted that labelled diacylglycerophosphomonomethylethanolamines, which are intermediates in the biosynthesis of diacylglycerophosphocholines from diacylglycerophosphoethanolamines[37], can be detected in soybean cultures that have been incubated with labelled ethanolamine[8].

It is often believed that acetate is incorporated mainly into the acyl moieties of phospholipids, whereas ethanolamine, choline, serine and inositol are incorporated as such or after they have been converted into other bases. Our results on soybean cell cultures, as presented in Table 4, show that, when the cells are incubated with [1-^{14}C]acetate, as expected the radioactivity appears mainly in the acyl moieties of diacylglycerophosphoethanolamines and diacylglycerophosphocholines.

Table 4. Incorporation of [1-^{14}C]acetate into acyl and nonacyl moieties of phospholipids in soybean cell cultures[34]

Phospholipids	Acyl moieties	Nonacyl-moieties
Diacylglycerophosphoethanolamines	92.6	7.4
Diacylglycerophosphocholines	74.2	25.8

Values are expressed in % of radioactivity in intact phospholipids
9 day cells grown in the dark were incubated with the radioactive substrate for 48 h

When [2-^{14}C]ethanolamine is supplied as precursor, radioactivity is incorporated into the acyl moieties as well as the rest of the molecules of both diacylglycerophosphoethanolamines and diacylglycerophosphocholines. Cultures that have been incubated with [methyl-^{14}C]choline contain diacylglycerophosphocholines labelled in the base as well as diacylglycerophosphoethanolamines labelled in the acyl moieties and, to a lesser degree, in the base. A major portion of the radioactivity of [3-^{14}C]serine appears in the nonacyl moieties of diacylglycerophosphocholines and diacylglycerophosphoethanolamines as shown in Table 5.

These results indicate that different bases of phospholipids are not only interconvertable[8, 37], but that they are also to some extent degraded to acetate and then incorporated into acyl moieties of phospholipids and other lipid classes. This is confirmed by the finding that radioactivity from labelled bases appears also in acyl moieties of triacylglycerols and in the galactolipids[34].

The radioactivity from ^{14}C-labelled precursors that is incorporated into constituent fatty acids of phospholipids of soybean cell cultures appears mainly in palmitic, oleic, linoleic and linolenic acids, though in various levels depending on the duration of incubation of the cells with the precursor. After only short periods of incubation with ^{14}C-labelled acetate, radioactivity is concentrated mainly in oleic and palmitic

Table 5. Incorporation of ^{14}C-labelled precursors into acyl and nonacyl moieties of phospholipids in soybean cell cultures[34]

Phospholipids	[2-^{14}C] Ethanolamine		[Methyl-^{14}C] Choline		[3-^{14}C] Serine	
	Acyl moieties	Nonacyl moieties	Acyl moieties	Nonacyl moieties	Acyl moieties	Nonacyl moieties
Diacylglycerophospho-ethanolamines	66.2	33.8	79.3	20.7	15.9	84.1
Diacylglycerophospho-cholines	54.3	45.7	8.7	91.3	13.4	86.6

Values are expressed in % of radioactivity in intact phospholipids
9 day cells grown in the dark were incubated with the radioactive substrates for 48 h

acids of diacylglycerophosphoethanolamines and diacylglycerophosphocholines. After incubation up to 22 h, the radioactivity in linoleic and linolenic acids increases at the expense of radioactivity in oleic acid; the activity in palmitic acid remains about constant[33]. We have found that soybean cultures which were incubated for 48 h with [2-^{14}C]ethanolamine, [methyl-^{14}C]choline or [3-^{14}C]serine contained only small proportions of labelled palmitic acid but, in most cases, major amounts of labelled oleic acid in their phospholipids[34]. These patterns are similar to the pattern reported for labelled fatty acids of total lipids from soybean cell cultures that have been incubated with [1-^{14}C]acetate for 5 h[13].

The mechanisms of biosynthesis of fatty acids in plant cell cultures are discussed later in this article.

3.1.3 Galactolipids

As mentioned earlier, heterotrophic cell cultures produce only very low proportions of monogalactosyldiacylglycerols and digalactosyldiacylglycerols[1,9]. For example, 7 day cultures of soybean contain, in their total lipids, only 0.5 to 0.9% galactose, whereas 21 day cultures contain even much smaller proportions, 0.05–0.2%[35]. It is possible to stimulate the biosynthesis of galactolipids in cell cultures by rendering them photosynthetically active[3-5]. Table 6 shows that photoautotrophic cell cultures of *Chenopodium rubrum*, in contrast to heterotrophic cultures, contain large proportions of monogalactosyldiacylglycerols and digalactosyldiacylglycerols. These classes of galactolipids are known to predominate in lipids from photosynthetically active plants and plant organs[27,28,38].

Resembling galactolipids from photosynthetically active plants and plant organs[27] and chloroplasts[38], galactolipids from photomixotrophic[3,4] and photoautotrophic[5] plant cell cultures contain relatively high levels of linolenic acid. So far, it is not known whether linolenic acid is synthesized *via* desaturation of fatty acids in intact galactolipid molecules, or whether it is introduced into galactolipids through acyl exchange.

Studies on the incorporation of radioactively labelled precursors into galactolipids of cell cultures have been made, unfortunately, on heterotrophic cultures only[29,33-35]. Interesting results can be expected when studies are repeated on

Table 6. Proportions of galactolipids in total lipids from photoautotrophic and heterotrophic cell cultures of *Chenopodium rubrum*[5]

Galactolipid	Photoautotrophic cultures* (Duration of incubation in days)			Heterotrophic cultures** (Duration of incubation in days)
	(7)	(14)	(21)	(21)
Monogalactosyl-diacylglycerols	9.1	10.8	5.2	0.2
Digalactosyl-diacylglycerols	5.7	8.0	5.1	0.1

Values are expressed in mg/g dry cells
* Grown under continuous illumination
** Grown in the dark

photosynthetically active cell cultures. In one study involving ^{32}P- and $^{35}SO_4$-incorporation into callus cultures of *Kalenchoë crenata* it was shown that diacylglycerophosphoglycerols and sulfoquinovosyldiacylglycerols increased during greening of the cultures[39]. In addition, it was found that more galactolipids were synthesized[39].

When heterotrophic cell cultures of soybean and carrot are incubated with ^{14}C-acetate[33, 34] and [2-^{3}H]glycerol[29], respectively, only about 8% of the radioactivity in total lipids are incorporated into galactolipids. It has been reported that during prolonged incubation with ^{14}C-labelled acetate, more activity appears in these lipids[33]. It can be expected that much larger proportions of radioactivity will be incorporated into the galactolipids of photosynthetically active cultures.

[1-^{14}C]Acetate is incorporated into palmitic, oleic, linoleic and linolenic acids of monogalactosyldiacylglycerols and digalactosyldiacylglycerols[33]. The major portion of radioactivity appears in oleic and palmitic acids. It is to be expected that the pattern of labelling of acyl moieties in galactolipids will be different when photosynthetically active cultures are investigated instead of heterotrophic cultures.

[1-^{3}H]Galactose is incorporated into lipids of heterotrophic soybean cultures; it appears both in monogalactosyldiacylglycerols and digalactosyldiacylglycerols[35].

We suggest that mixotrophic and photoautotrophic cell cultures be used instead of heterotrophic cultures for studies concerned with the metabolism of galactolipids and their constituent fatty acids. In this context, it is to be mentioned that heterotrophic cultures lack sulfoquinovosyldiacylglyerols whereas photoautotrophic cultures synthesize appreciable amounts of this lipid class, about 1.5 to 5% of total lipids[5].

3.1.4 Fatty Acids

Reference has been made to the incorporation of ^{14}C-labelled acetate into the acyl moieties of various lipid classes of plant cell cultures. Other metabolic reactions involving fatty acids include elongation and desaturation. Studies on the biosynthesis of cyclic fatty acids deserve special consideration. In addition, the synthesis of fatty acids in intact proplastids, that have been isolated from plant cell cultures, should be discussed.

3.1.4.1 Elongation and Desaturation of Straight — Chain Fatty Acids

Lipids in plant cell cultures contain various proportions of very long-chain fatty acids with even and odd numbers of carbon atoms up to 26[6]. The lipids of callus cultures usually contain higher proportions of these very long-chain fatty acids than those of suspension cultures[40]. So far, there are no data available on the mechanism of fatty acid elongation in plant cell cultures. When ^{14}C-labelled acetate is used as substrate by soybean cultures, 5% of the activity in the fatty acids is incorporated into icosanoic acid, 20:0[13]. Supposedly[13], in cell cultures of soybean there is an ACP dependent *de novo* sequential fatty acid synthesis similar to that in intact plants[41]. Acetate is converted to palmitoyl-ACP, and then a specific elongation system elongates palmitoyl-ACP to stearoyl-ACP. Finally, a specific stearoyl-ACP desaturase converts stearoyl-ACP to oleoyl-ACP[42]. Reportedly[13], free palmitic and stearic acids do not enter the acyl-ACP pool and are neither elongated nor desaturated because they cannot be converted to acyl-ACP derivatives.

Cell cultures of soybean were found to absorb rapidly, not only ^{14}C-labelled acetate, but also palmitic, stearic, oleic, and linoleic acids. The cells incorporate these fatty acids into various lipid classes[13]. Fig. 4 illustrates the uptake and metabolism of [1-^{14}C]stearic, [1-^{14}C]oleic and [1-^{14}C]linoleic acids by the cells as a function of time.

Within minutes after addition to the medium, considerable proportions of exogenous fatty acids appear inside the cells where they are incorporated mainly into diacylglycerophosphocholines and triacylglycerols and, to a lesser extent, into diacylglycerophosphoethanolamines. Saturated fatty acids, such as palmitic and stearic acids, are not metabolized further during 5 h of incubation (Table 7). In contrast, unsaturated fatty acids, such as oleic acid and, to a much lesser extent, linoleic acid, are desaturated to linoleic and linolenic acids, respectively. When the incubation period is extended to 22 h[33], rather high levels of labelled linolenic acid are formed in the lipids of soybean cell cultures.

Table 7. Incorporation of ^{14}C-labelled fatty acids into soybean cell cultures[13]

Chain length: Number of double bonds	[1-^{14}C] Palmitic acid	[1-^{14}C] Stearic acid	[1-^{14}C] Oleic acid	[1-^{14}C] Linoleic acid
16:0	100	0	0	0
18:0	0	100	0	0
18:1	0	0	83	0
18:2	0	0	17	95
18:3	0	0	0	<5

Values are expressed in % of radioactivity
5 days old cells were incubated with the labelled fatty acids for 5 h

It is believed that oleic acid is first activated to a CoA thioester, and then desaturated to linoleoyl-CoA; both thioesters are then transferred to triacylglycerols, diacylglycerophosphoethanolamines and diacylglycerophosphocholines. Fig. 5 shows

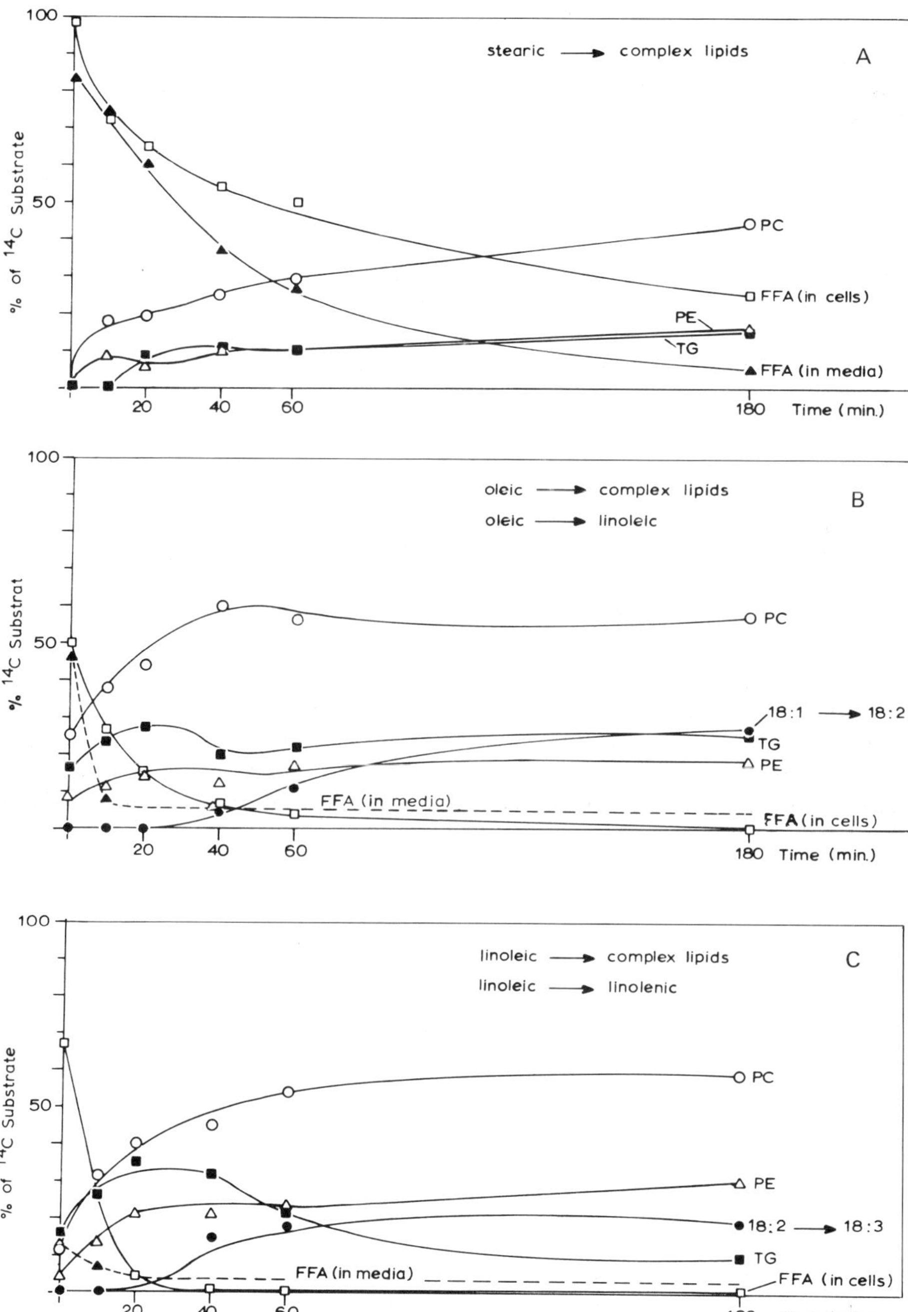

Fig. 4 Uptake and metabolism of [1-^{14}C]stearic acid (A), [1-^{14}C]oleic acid (B) and [1-^{14}C]linoleic acid (C) by cell cultures of soybean as a function of time [13]
The curves depict the course of incorporation of the substrates into TG: Triacylglycerols; PE: Diacylglycerophosphoethanolamines; PC: Diacylglycerophosphocholines; FFA: Free fatty acids in media and in cells

a scheme[33] illustrating a proposed mechanism of desaturation of fatty acids in soybean cell cultures.

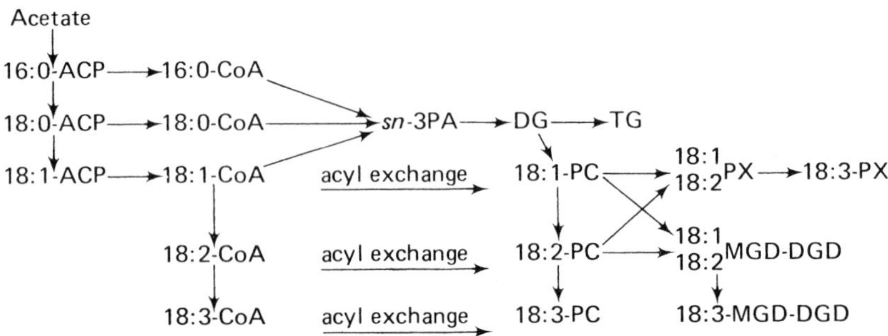

Fig. 5 Scheme for proposed mechanism of desaturation sequence of fatty acids in cell cultures of soybean[33]

In soybean cell cultures, as in intact plants, both the desaturation of oleic to linoleic, and of linoleic to linolenic acids, is dependent on the availability of molecular oxygen[13].

In this context, it is of interest to note that soybean cell cultures can incorporate positional isomers of *cis*- and *trans*-octadecenoic acids in their acyl lipids[43]. Each of positional isomers ranging from Δ8- to Δ15-*cis*-octadecenoic acids and Δ7- to Δ16-*trans*-octadecenoic acids is incorporated into acyl lipids; the Δ9-*cis*- and the Δ9-*trans*-isomers are the preferred substrates. No *cis* to *trans* isomerization occurs during incorporation of fatty acids into acyl lipids of cell cultures. Only the Δ9-*cis*-isomer (oleic acid) is preferentially incorporated into position 2 of diacylglycerophosphocholines, diacylglycerophosphoethanolamines and triacylglycerols, whereas all the other isomers excluding the Δ9-*trans*-octadecenoic acid (elaidic acid) exhibit affinity for position 1 of glycerophospholipids and positions 1 and 3 of triacylglycerols[44].

3.1.4.2 *Biosynthesis of Cyclic Fatty Acids*

The lipids in callus cultures of *Malva sp.* contain substantial proportions of both cyclopropane and cyclopropene fatty acids[21]. In contrast to the seeds of plants belonging to the family *Flacourtiaceae*, tissue cultures of these plants contain only small proportions of cyclopentenyl fatty acids in their lipids[22]. Repeated subculturing of the cells results in lowering the amounts of these fatty acids even further, until there are only traces in the total fatty acids[45].

The pathways of biosynthesis of cyclopropane and cyclopropene fatty acids have been investigated in tissue cultures[46]. By incubating cultures of *Malva sp.* with labelled acetate and methionine, it was confirmed that the methylene group in the ring originates from the methyl groups of methionine. It has been suggested that dihydrosterculic acid is formed initially. It is then desaturated to sterculic acid and/or α-oxidized to malvalic and dihydromalvalic acids. The validity of this assumption was confirmed by the fact that labelled oleic acid is converted to dihydrosterculic and sterculic acids and that labelled dihydrosterculic acid can be desaturated to sterculic acid.

The mechanism of biosynthesis of cyclopentenyl fatty acids has been studied in cell cultures of *Idesia polycarpa*[45,47]. Such cultures can incorporate labelled aleprolic acid (cyclopentenyl carbonic acid) and convert it to long-chain cyclopentenyl fatty acids. The chief site of synthesis of these acids in cell cultures is the cytosol although some biosynthetic activity can be detected in the proplastid[47] as well. Cell cultures of the same species can incorporate [1,2-^{14}C] α-ketopimelic acid into cyclopentenyl fatty acids, however, this substrate is decomposed in the cell and radioactivity appears not only in the cyclopentenyl fatty acids but also in straight-chain fatty acids as well as in free amino acids[45]. The incorporation of radioactivity from α-ketopimelic acid into cyclopentenyl fatty acids is considerably higher than from acetate (Table 8).

Table 8. Incorporation of labelled substrates in cyclopentenyl fatty acids and straight-chain fatty acids of *Idesia polycarpa* cell cultures[45]

Fatty acids	[1,2-^{14}C]α-Keto-pimelic acid*	[1-^{14}C] Acetate*	Endogenous fatty acid**
$C_{14:0}$, $C_{15:0}$	5.9	3.1	0.7
$C_{16:0}$	28.4	44.5	30.1
Hydnocarpic ($C_{16:1cy}$)	5.1	0.3	Trace
$C_{17:0}$, $C_{18:0}$	54.4	50.5	69.2
Chaulmoogric ($C_{18:1cy}$)	6.2	0.5	—

* Values are expressed in % of radioactivity in fatty acids
** Values are expressed in wt.%

3.1.4.3 Biosynthesis of Fatty Acids in Intact Proplastids

In cells of intact plants, desaturation of oleic to linoleic acid is believed to be associated with the endoplasmic reticulum, whereas, the desaturation of linoleic to linolenic acid seems to occur in the chloroplasts[48,49]. Heterotrophic, photomixotrophic[3,4,50-54] and photoautotrophic[5,55] cultures are suitable for studies concerned with the localization of the sites of fatty acid biosynthesis in the cell.

In the available literature, there is only a single publication which describes the biosynthesis of fatty acids in proplastids isolated from heterotrophic cultures of soybean[56]. The cells were homogenized and, by differential centrifugation, subcellular fractions were isolated and incubated separately with labelled precursors of fatty acids. Figure 6 shows an electron micrograph of a proplastid from a cell culture and Table 9 presents data on the incorporation of ^{14}C-labelled substrates into the total fatty acids from proplastids.

It is apparent that, in intact proplastids, [1-^{14}C]acetate is incorporated best. Good incorporation is observed also, when [3-^{14}C]pyruvate serves as substrate; but when [1-^{14}C] pyruvate is used, incorporation of radioactivity cannot be observed. This result indicates that proplastids from soybean cell cultures contain pyruvate dehydrogenase which converts [3-^{14}C]pyruvate to [2-^{14}C]acetyl-CoA and CO_2, and converts [1-^{14}C]pyruvate to acetyl-CoA and $^{14}CO_2$. Table 10 shows the incorpora-

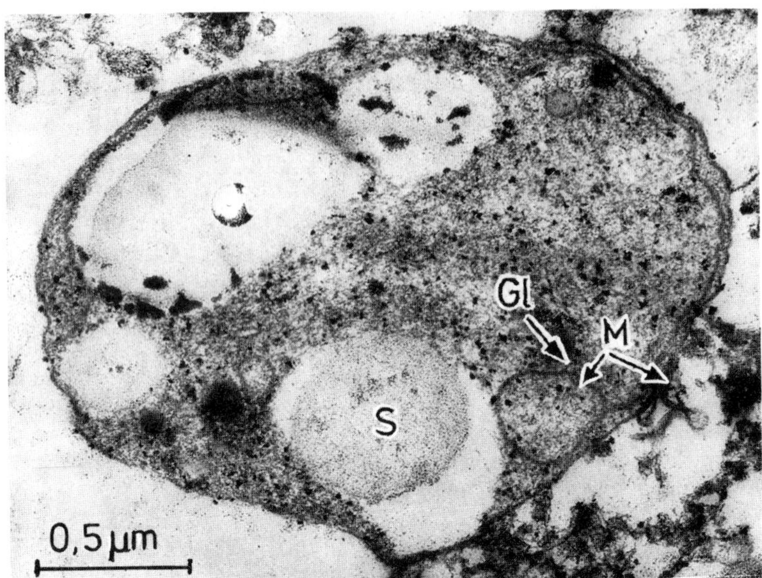

Fig. 6 Electron micrograph of an intact proplastid from a cell culture of soybean [56]
S: Starch; Gl: Osmiophilic droplets (plastoglobuli); M: Membranes

Table 9. Incorporation of ^{14}C-labelled substrate into fatty acids of proplastids from soybean cell cultures[56)]

	Intact proplastids			Disrupted proplastids
	[1-^{14}C] Acetate	[2-^{14}C] Malonate	[3-^{14}C] Pyruvate	[2-^{14}C] Malonyl CoA
pMol substrate/	2988	1448	1432	568
mg protein/3 h	2978	1642		440

7 day cells grown in the dark were used

Table 10. Incorporation of ^{14}C-labelled substrates into fatty acids of proplastids from soybean cell cultures[56)]

Chain lenght: Number of double bonds	Intact proplastids			Disrupted proplastids
	[1-^{14}C] Acetate	[2-^{14}C] Malonate	[3-^{14}C] Pyruvate	[2-^{14}C] Malonyl CoA
<16	0.7	3.0	3.8	8.3
16:0	27.5	22.2	10.2	13.6
16:1	0.3	2.3	7.0	3.3
18:0	44.9	53.3	43.5	42.4
18:1	23.1	13.4	28.0	18.5
18:2	0.7	0.6	3.8	3.7
18:3	1.5	1.4	2.0	5.9
20:0	1.4	3.8	1.7	4.4

Values are presented in % ^{14}C
7 day cells grown in the dark were used

tion of ^{14}C-labelled substrates into the different fatty acids of proplastids from soybean cell cultures.

The fatty acid synthetase activity of *Idesia polycarpa*[47] and soybean[56] cell cultures is restricted to the proplastids. In intact nongreen plant tissues such as developing castorbean seeds[57–60], avocado and cauliflower buds[61, 62], proplastids are known to be the site of synthetase activity. In green leaves, however, chloroplasts are the site of fatty acid synthesis[63].

In proplastids from soybean cell cultures, maximal activity of the fatty acid synthesizing system is at pH 8.0–8.2, with acetate as substrate, whereas in those from *Idesia polycarpa* the optimum pH-value is 5.6[47, 56].

3.1.5 Sterols and Steroidal Compounds

In the late stationary phase of growth, plant cell cultures contain relatively high proportions of sterols and complex lipids containing sterols[9, 64]. ^{14}C-labelled acetate has been shown to be incorporated into sterols, steryl esters, steryl glycosides and acyl steryl glycosides from cultures of soybean[33, 34] and carrot[29]. Cell cultures synthesize the same sterols that are produced in intact plants, *i.e.*, mainly β-sitosterol and stigmasterol[9]. In addition, small proportions of less common sterols[65–68] are also present. In a few studies, the absorption and metabolism of supplemented cholesterol by plant cell cultures have been reported[14, 15].

The synthesis of industrially valuable steroidal compounds in plant cell cultures has received considerable interest. Thus, steroidal glycoalkaloids, which occur in intact plants belonging to the genus *Solanum*[69–72] and which are suitable as substitutes for diosgenin in the synthesis of steroidal hormones[73], were detected in cell cultures of *Solanum sp.*[74–76], though at rather low levels. The influence of different plant hormones, singly, and in various combinations, on cultures of *Solanum sp.* was found to be manifested in some changes in steroidal contents of the cells, indicating a possibility of biochemical regulation by auxins[77]. Recently, it has been found that the stimulation of organ differentiation in cell cultures of *Solanum sp.* is associated with the promotion of biosynthesis of steroidal glycoalkaloids in these cultures[78, 79].

For more details on the sterols in plant cell cultures and their metabolism, the reader is referred to several other chapters in this volume.

3.2 Degradation of Lipids

In studies concerned with the metabolism of lipids in plant cell cultures, major emphasis has been placed on anabolic reactions. So far, catabolic processes have received only limited attention. Within the coming years, after extensive information on the biosynthesis of lipids will have accumulated, more effort will be devoted to the problem of degradation of lipids in cell cultures.

3.2.1 Lipolytic Enzymes

It appears that plant cell cultures, as intact plant tissues, contain the whole complement of enzymes that catalyze the decomposition of various lipid classes. Experimental evidence which confirms the validity of this statement is available in only a very few cases. Thus, it has been demonstrated that several strains of cell cultures of

Daucus carota exhibit phospholipase D activity[29]. This enzyme catalyzes the hydrolytic cleavage of the bases from phospholipids producing phosphatidic acid; it occurs commonly in intact plants[30].

3.2.2 Turnover of Lipid Classes

There are a few rather interesting results on the turnover of some lipid classes in plant cell cultures. Glycerol was selected for turnover measurements of lipid classes in cell cultures of *Daucus carota*[29].

Cells were incubated with [2-^{3}H]glycerol for 3 days, and then were inoculated into fresh medium containing unlabelled glycerol; cell samples were analyzed at 12 h intervals. It was found that diacylglycerophosphocholines, diacylglycerophosphoinositols, monogalactosyldiacylglycerols, and digalactosyldiacylglycerols have half-life times of about 45 h each, *i.e.*, in the order of half of a generation time of the cells. The half-life time of diacylglycerophosphoethanolamines is somewhat shorter, whereas that of triacylglycerols is longer. It is surprising that triacylglycerols which are considered as storage lipids, exhibit a turnover rate that is below the generation time of the cells[29].

Similar pulse-chase experiments have been done on soybean cell cultures[8] and the turnover rates of phospholipids were measured. Labelled substrates such as [U-^{14}C]ethanolamine, [1,2-^{14}C]choline, [3-^{14}C]serine, [2-^{14}C]acetate, and [U-^{14}C]glycerol were used. The half-life times of individual phospholipids, synthesized from the labelled substrates in the cells, are summarized in Table 11.

Table 11. Turnover rates of phospholipids in soybean cell cultures[8]

Phospholipid	[U-^{14}C] Ethanolamine	[1,2-^{14}C] Choline	[3-^{14}C] Serine
Diacylglycerophosphoethanolamines	136		300
Diacylglycerophosphocholines	92	36–96	99
Diacylglycerophosphoserines			54

Values are half-life times in h
6 day cells were used

The results indicate that the turnover rates of phospholipids in plant cell cultures are generally similar to those of phospholipids in intact plant and animal organs[81, 82]. The individual phospholipids in cell cultures vary in their turnover rates. In addition, there appears to be considerable variation also in the turnover rates of different portions of molecules of the same phospholipid, for example the glycerol backbone of diacylglycerophosphoethanolamines and diacylglycerophosphocholines as compared with the acyl moieties and base groups. Variations in the turnover rates of portions of phospholipid molecules are known to occur in animal systems[83].

The turnover rates of phospholipids in cell cultures are dependent on the growth phase. The half-life times of diacylglycerophosphocholines in soybean cell cultures, for example, were found to be 36 and 96 h during the exponential and the stationary growth phases, respectively[8]. The turnover rates of different lipid classes in cell cultures can certainly be influenced by changing the environmental conditions prevailing during growth.

The degradation products of diacylglycerophosphoserines and diacylglycerophosphoethanolamines are probably utilized for the synthesis of diacylglycerophosphocholines in plant cell cultures[8] (see also Table 11). Neither in plants nor in animals could satisfactory evidence be obtained for the *in vivo* operation of such pathways[84]. These facts demonstrate clearly the value of plant cell culture techniques in studies on lipid metabolism.

4 Plant Cell Cultures as a Tool in Lipid Research

For lipid research, plant cell cultures provide several advantages over intact plants:

a) — Cultures are available for study throughout the year and practically indefinitely, irrespective of the geographic distribution of the mother plant and the climatic conditions.

b) — The life cycles of the cell cultures do not exceed 10–15 days, whereas those of intact plants extend for months or even years.

c) — The handling of the cultures, for example, their storage and extraction, is easier than that of plant organs.

d) — Culture cells respond to changes in environmental conditions more rapidly and more sensitively than intact plants and hence, lipid metabolism in cultures can be restricted easily to pathways suitable for the study.

e) — Protoplasts can be produced from cell cultures more easily than from intact plants; they are valuable because of their lack of cell walls, although they contain all cell organelles, intact and functional.

We have referred to several examples in which plant cell cultures have been successfully used in lipid research.

It has been shown that through the cell culture technique, light is being shed on metabolic pathways of lipids in various biological systems.

In view of the fact that cell cultures show patterns of incorporation of labelled acetate into cell fatty acids similar to those of seed slices[32], they have frequently been described as "model systems". This term can be misleading. Cell cultures are certainly valuable tools for research, but they are, by no means, identical with the intact plants and plant organs in metabolic activity. In addition, cell cultures are not fixed entities with fixed activities irrespective of age or cultural conditions. Their value lies in that they respond sensitively in their metabolism to the environment. Again, the activities of cells in the exponential phase of growth are not identical with those of cells in the stationary phase. Further, heterotrophic cultures are metabolically not identical with photomixotropic or photoautotrophic cultures of the same plant. This diversity in the metabolic activities of cell cultures should receive particular attention when they are to be used as "model systems".

We see in this diversity very valuable advantages for lipid research. Heterotrophic cultures show a certain similarity to plant roots in their lipid metabolism, whereas photomixotrophic and photoautotrophic cultures are rather similar to green leaves and other green organs. Cultures in the exponential phase of growth are similar to the actively dividing meristematic tissues of intact plants, whereas cells in the stationary phase resemble adult tissues. Cultures which have started embryogenesis are metabolically similar to intact embryos and seeds. Yet, these similarities are general, and specific differences should always be expected. Although, for example, photomixotrophic cultures of *Nicotiana tabacum* resemble leaves in that they produce relatively large proportions of monogalactosyldiacylglycerols and digalactosyldiacylglycerols with high levels of constituent linolenic acid, they differ in that their lipids lack *trans*-3-hexadecenoic acid and hexadecatrienoic acid, which occur in considerable amounts in lipids of tobacco leaves. However, it will certainly be possible to promote the synthesis of these fatty acids in green cell cultures since this will depend only on finding the specific experimental conditions.

Plant cell cultures are valuable in applied lipid research. Thus, it is possible to prepare phospholipids labelled in their base moiety and/or in their acyl moieties by supplying cell cultures with appropriate precursors. Good yields of 1.2-di-*O*-acyl-glycerophospho-[2-^{14}C]ethanolamines and 1.2-di-*O*-acylglycerophospho-[methyl-^{14}C]-cholines have been isolated from rape and soybean cultures that were incubated with ^{14}C-labelled precursors for a few minutes[34]. Randomly labelled lipids have been obtained by incubating cell cultures of rape and soybean with [1-^{14} C]acetic acid, and uniformly labelled lipids have been isolated from cultures that were incubated with a mixture of [1-^{14}C]acetic acid plus [2-^{14}C]acetic acid[85].

It can be expected that, when photoautotrophic cultures are grown in ^{14}C-labelled carbon dioxide as the sole source of carbon, uniformly labelled lipids will also be produced.

Plant cell cultures should be particularly suitable for the preparation of phospholipids labelled with stable isotopes, such as the heteroatoms ^{15}N and ^{31}P. Moreover, such cultures may be of value in producing complex lipids whose acyl moieties contain a spin radical, a fluorescent group, or a light-sensitive label. Yet, whether fatty acids labelled with such entities indeed get incorporated into the acyl lipids of plant cell cultures remains to be investigated.

We suggest that efforts be devoted to the application of plant cell cultures in the production of arachidonic acid and its biotransformation to prostaglandins. For the production of arachidonic acid, cell cultures of mosses, cultivated under special conditions, appear to be most promising. The biosynthesis of prostaglandins in animal cell cultures is receiving considerable attention[86].

Various enzymes involved in lipid metabolism can be isolated from plant cell cultures. Reference has been made to the occurrence of phospholipase D in cell cultures of *Daucus carota*[29]. Of special interest is that "cholesterol oxidase" an enzyme used for the estimation of cholesterol in blood serum has been shown to occur in cell cultures of *Brassica napus* and *Glycine max*[15]. Procedures for the isolation of such enzymes should be evolved. To our knowledge, only one enzyme, tobacco acid pyrophosphatase (TAP)[87] is being commercially produced from plant cell cultures.

Detailed knowledge of the influence of growth phase, nutrient medium, and environmental conditions on the lipids of plant cell cultures can be exploited in programming such cultures for the production of valuable compounds[12]. By controlling these factors, it is possible to modify the properties of the cells' organelles and membranes[88]. This may provide a tool for facilitating the excretion and recovery of cell contents and/or the process of protoplast fusion. In this context, it should be mentioned that the chemical modification of plant membranes has been achieved in intact seedlings and tuber slices[89, 90]. Recently, the molecular species of phospholipids in plant cell cultures have been determined[91, 92], and the influence of membrane fluidity on protoplast fusion as effected by the phase transition of these phospholipids has been discussed[91].

Cell culture techniques have become of practical use in the vegetative propagation of jojoba (*Simmondsia chinensis*), a desert shrub that is difficult to propagate by conventional means. The seeds of this plant contain about 50% of a liquid wax having great commercial value. Several laboratories are engaged in studying the route of biosynthesis of wax esters in cell cultures of jojoba. Others are concerned with the effect of embryogenesis on the fatty acid composition of lipids in cell cultures[31, 93–95].

It has been established that plant cell cultures respond to chilling in a manner similar to that of organized plant tissues[12, 90]. Hence, such cultures have been suggested as a tool for studies of the chilling phenomenon and the selection of chillingresistant cell lines[90].

Acknowledgment:

The authors' work has been supported by the 'Bundesministerium für Forschung und Technologie', D-5300 Bonn.

5 References

1. Radwan, S. S., Mangold, H. K.: Adv. Lipid Res. *14*, 171 (1976)
2. Mangold, H. K.: *In*: Plant Tissue and Cell Culture and its Bio-Technological Application. Barz, W., Reinhard, E., Zenk, M. H. (eds.), p. 55. Berlin: Springer 1977
3. Siebertz, H. P., Heinz, E., Bergmann, L.: Plant Sci. Lett. *12*, 119 (1978)
4. Radwan, S. S., Mangold, H. K., Hüsemann, W., Barz, W.: Chem. Phys. Lipids *24*, 79 (1979)
5. Hüsemann, W., Radwan, S. S., Mangold, H. K., Barz, W.: Planta (in press)
6. Radwan, S. S., Mangold, H. K., Spener, F.: Chem. Phys. Lipids *13*, 103 (1974)
7. De Silva, N. S., Fowler, M.: Phytochemistry *15*, 1735 (1976)
8. Moore, Jr., T. S.: Plant Physiol. *60*, 754 (1977)
9. Radwan, S. S., Spener, F., Mangold, H. K., Staba, E. J.: Chem. Phys. Lipids *14*, 72 (1975)
10. Radwan, S. S.: Fette, Seifen, Anstrichm. *77*, 181 (1975)
11. Radwan, S. S., Mangold, H. K.: Chem. Phys. Lipids *14*, 87 (1975)
12. Radwan, S. S., Grosse-Oetringhaus, S., Mangold, H. K.: Chem. Phys. Lipids *22*, 177 (1978)
13. Stumpf, P. K., Weber, N.: Lipids *12*, 120 (1977)
14. Heble, M. R., Narayanaswamy, S., Chadha, M. S.: Phytochemistry *15*, 1911 (1976)
15. Weber, N.: Z. Pflanzenphysiol. *87*, 355 (1978)
16. Radwan, S. S.: Phytochemistry *15*, 727 (1976)
17. Shin, B. S., Mangold, H. K., Staba, E. J.: *In*: Les Cultures de Tissus de Plantes. Colloque No. 193, p. 51. Paris CNRS 1971
18. Staba, E. J., Shin, B. S., Mangold, H. K.: Chem. Phys. Lipids *6*, 291 (1971)
19. Gemmrich, A. R.: unpublished
20. Ellenbracht, F., Barz, W., Mangold, H. K.: in preparation
21. Yano, I., Nichols, B. W., Morris, L. J., James, A. T.: Lipids *7*, 30 (1972)
22. Spener, F., Staba, E. J., Mangold, H. K.: Chem. Phys. Lipids *12*, 344 (1974)

23. Spener, F., Mangold, H. K.: Phytochemistry *14*, 1369 (1975)
24. Spener, F., Mangold, H. K.: Biochemistry *13*, 2241 (1974)
25. Hüsemann, W., Barz, W.: Physiol. Plant *40*, 77 (1977)
26. Kates, M., Wilson, A. C., de la Roche, A. I.: *In*: Advances in the Biochemistry and Physiology of Plant Lipids. Appelqvist, L.-Å., Liljenberg, C. (eds), p. 329. Amsterdam, Elsevier/North Holland 1979
27. Kates, M.: Adv. Lipid Res. *8*, 225 (1970)
28. Radwan, S. S., Mangold, H. K., Barz, W., Hüsemann, W.: Chem. Phys. Lipids *25*, 101 (1979)
29. Kleinig, H., Kopp, C.: Planta *139*, 61 (1978)
30. Kleinig, H., Steinki, C., Kopp, C., Zaar, K.: Planta *140*, 233 (1978)
31. Jones, L. H.: *In*: Industrial Aspects of Biochemistry. Spencer, B. (ed.), Vol. XXX, p. 813. Amsterdam: North Holland 1974
32. Stearns, Jr., E. M., Morton, W. T.: Lipids *10*, 597 (1975)
33. Wilson, A. C., Kates, M., de la Roche, A. I.: Lipids *13*, 504 (1978)
34. Mangold, H. K., Radwan, S. S.: *In*: Plant Cell Cultures: Results and Perspectives, Ciferri, O., Paris, B. (eds.), Amsterdam, Elsevier/North Holland, 1979
35. Sabinski, F., Fromme, H. G., Spener, F.: unpublished
36. Gregor, H. D.: Chem. Phys. Lipids *20*, 77 (1977)
37. Moore, Jr., T. S.: Plant Physiol *57*, 383 (1976)
38. Hitchcock, C., Nichols, B. W.: Plant Lipid Biochemistry, p. 68. London: Academic Press 1971
39. Thomas, D. R., Stobart, A. K.: J. Exp. Bot. *21*, 274 (1970)
40. Gemmrich, A. R., Schraudolf, H.: Chem. Phys. Lipids (in press)
41. Jaworski, J. G., Goldschmidt, E. E., Stumpf, P. K.: Arch. Biochem. Biophys. *163*, 769 (1974)
42. Jaworski, J. G., Stumpf, P. K.: Arch. Biochem. Biophys. 162, 158 (1974)
43. Richter, I., Weber, N., Mangold, H. K., Mukherjee, K. D.: Z. Naturforsch. *33c*, 303 (1978)
44. Weber, N., Richter, I., Mangold, H. K., Mukherjee, K. D.: Planta *145*, 479 (1979)
45. Tober, I., Spener, F.: unpublished
46. Yano, I., Morris, L. J., Nichols, B. W., James, A. T.: Lipids *7*, 35 (1972)
47. Buchholz, S., Spener, F.: unpublished
48. Slack, C. R., Roughan, P. G.: Biochem. J. *152*, 217 (1975)
49. Tremolieres, A., Mazliak, P.: Plant Sci. Lett. *2*, 193 (1974)
50. Bergmann, L., Bälz, A.: Planta *70*, 285 (1966)
51. Hanson, A. D., Edelman, J.: Planta *102*, 11 (1972)
52. Neumann, K.-H., Saafat, A.: Plant Physiol. *51*, 685 (1973)
53. Nato, A., Bazetoux, S., Mathieu, Y.: Physiol. Plant. *41*, 116 (1977)
54. Yamaya, T., Ojima, K., Ohira, K.: Soil Sci. Plant Nutr. *23*, 59 (1977)
55. Yamada, Y., Sato, F., Hagimori, M.: *In*: Frontiers of Plant Tissue Culture 1978, T. A. Thorpe (ed.), p. 453. Calgary: University of Calgary Press 1978
56. Nothelfer, H. G., Barckhaus, R. H., Spener, F.: Biochim. Biophys. Acta *489*, 370 (1977)
57. Zilkey, B. F., Canvin, D. T.: Can. J. Bot. *50*, 323 (1972)
58. Yamada, M., Usami, Q.: Plant Cell Physiol. *16*, 879 (1975)
59. Drennan, C. H., Canvin, D. T.: Biochim. Biophys. Acta *187*, 193 (1969)
60. Nakamura, Y., Yamada, M.: Plant Cell Physiol. *15*, 37 (1974)
61. Weaire, P. J., Kekwick, R. G. O.: Biochem. J. *146*, 425 (1975)
62. Weaire, P. J., Kekwick, R. G. O.: Biochem. J. *146*, 439 (1975)
63. Stumpf, P. K.: *In*: Recent Advances in the Chemistry and Biochemistry of Plant Lipids. Galliard, T., Mercer, E. I. (eds.), p. 95. London: Academic Press 1975
64. Stohs, S. J., Rosenberg, H.: Lloydia *38*, 181 (1975)
65. Itokawa, H., Akasu, M., Fujita, M.: Chem. Pharm. Bull. *21*, 1386 (1973)
66. Lee, P. K., Carew, D. P., Rosazza, J.: Lloydia *35*, 150 (1972)
67. Weber, N.: Phytochemistry *16*, 1849 (1977)
68. Heble, M. R., Narayanaswamy, S., Chadha, M. S.: Phytochemistry *15*, 681 (1976)
69. Saini, A. D.: Curr. Sci. *35*, 600 (1966)
70. Panina, V. V., Pimenova, T. B.: Khim.-Farm. Zh. *6*, 46 (1972)
71. Eid, M. N. A., Ahmed, S. S., El-Antably, H. M. M.: Egypt. J. Hortic. *1*, 3 (1974)

72. Chopra, R. N., Nayar, S. L., Chopra, I. C.: Glossary of Indian Medicinal Plants, p. 230, New Delhi: C.S.I.R., 1956
73. Schreiber, K.: *In*: The Alkaloids. Manske, R. H. F. (ed.), p. 192. New York: Academic Press 1968
74. Heble, M. R., Narayanaswami, S., Chadha, M. S.: Naturwissenschaften *7*, 351 (1968)
75. Heble, M. R., Narayanaswami, S., Chadha, M. S.: Science *161*, 1145 (1968)
76. Kadkade, P. G., Madrid, T. R.: Naturwissenschaften *64*, 147 (1977)
77. Heble, M. R., Narayanaswami, S., Chadha, M. S.: Phytochemistry *10*, 2393 (1971)
78. Uddin, A., Chaturved, H. C.: Planta Med. *37*, 90 (1979)
79. Kokate, C. K., Radwan, S. S.: Z. Naturforsch. *34c*, 634 (1979)
80. Wurster, C. F., Copenhaver, J. H.: Lipids *1*, 422 (1966)
81. Hayes, L. W., Jungaiwala, F. B.: Biochem. J.: *160*, 195 (1976)
82. Morré, D. J.: Annu. Rev. Plant Physiol. *26*, 441 (1975)
83. Omura, T., Siekevitz, P., Palade, G. E.: J. Biol. Chem. *242*, 2389 (1967)
84. Harwood, J. I.: Phytochemistry *15*, 1459 (1976)
85. Radwan, S. S., Mangold, H. K.: unpublished
86. Ohuchi, K., Levine, L.: J. Biol. Chem. *253*, 4783 (1978)
87. Efstratiadis, A., Vournakis, J. N., Keller, H. D., Chaconas, G., Dougal, D. K., Kafatos, F. C.: Nucleic Acid Res. *4*, 4165 (1977)
88. Sidorenko, P. G., Uvarov, G. A.: Tsitologya *14*, 1535 (1972)
89. Waring, A. J., Breidenbach, J. M., Lyons, J. M.: Biochim. Biophys. Acta *443*, 157 (1976)
90. Breidenbach, R. W., Waring, A. J.: Plant Physiol. *60*, 190 (1977)
91. Nishihara, M., Kito, M.: Biochim. Biophys. Acta *531*, 25 (1978)
92. Yamada, Y., Hara, Y., Senda, M., Nishihara, M., Kito, M.: Phytochemistry *18*, 423 (1979)
93. Warren, G. S., Fowler, M. W.: Planta *144*, 451 (1979)
94. de la Roche, A. I., Keller, W. A.: Z. Pflanzenzüchtg. *78*, 319 (1977)
95. Stringam, G. R.: Z. Pflanzenphysiol. *92*, 452 (1979)

Subject Index

Author Index Vol. 1–16

Advances in Biochemical Engineering

Managing Editor: A. Fiechter

"The series **Advances in Biochemical Engineering** is a very welcome addition to the literature, and... will contribute to the development, to quote the editors, of 'the yet to emerge hybrid discipline of biochemical engineering'." *Nature*

Volume 10

Immobilized Enzymes I

1978. 48 figures, 14 tables. VII, 177 pages
ISBN 3-540-08975-6

Contents:
W. H. Pitcher, jr.: Design and Operation of Immobilized Enzyme Reactors. *(120 ref.)*
S. A. Barker, P. J. Somers: *Biotechnology of Immobilized* Multienzyme Systems. *(64 ref.)*
R. A. Messing: Carriers for Immobilized Biological Active Systems. *(43 ref.)*
P. Brodelius: Industrial Applications of Immobilized Biocatalysts. *(206 ref.)*
B. Solomon: Starch Hydrolysis by Immobilized Enzymes. Industrial Applications. *(211 ref.)*

Volume 11

Microbiology, Theory and Application

1979. 76 figures, 35 tables. V, 180 pages
ISBN 3-540-08990-X

Contents:
D. Ramkrishna: Statistical Models of Cell Populations. *(61 ref.)*
S. Nagai: Mass and Energy Balances for Microbial Growth Kinetics. *(78 ref.)*
J. M. Scharer, M. Moo-Young: Methane Generation by Anaerobic Digestion of Cellulose-Containing Wastes. *(65 ref.)*
B. Metz, N. W. F. Kossen, J. C. van Suijdam: The Rheology of Modul Suspensions. *(65 ref.)*
M. Zlokarnik: Scale-up of Surface Aerators for Waste Water Treatment. *(22 ref.)*

Volume 12

Immobilized Enzymes II

1979. 99 figures, 64 tables. V, 253 pages
ISBN 3-540-09262-5

Contents:
H. M. Koplove, C. L. Cooney: Enzyme Production During Transient Growth. *(189 ref.)*
R. D. Schmid: Stabilized Soluble Enzymes. *(297 ref.)*
S.S. Wang, C.-K. King: The Use of Coenzymes in Biochemical Reactors. *(133 ref.)*
C. Wandrey, E. Flaschel: Process Development and Economic Aspects in Enzyme Engineering. Acylase L-methionine System. *(67 ref.)*
D. J. Graves, Y.-T. Wu: The Rational Design of Affinity Chromatography Separation Processes. *(34 ref.)*

Volume 13

Mass Transfer and Process Control

1979. 134 figures, 13 tables. VII, 214 pages
ISBN 3-540-09468-7

Contents:
W. Hampel: Applications of Microcomputers in the Study of Microbial Processes. *(97 ref.)*
Y. H. Lee, G. T. Tsao: Dissolved Oxygen Electrodes. *(150 ref.)*
H. Brauer: Power Consumption in Aerated Stirred Tank Reactor Systems. *(16 ref.)*
H. Blenke: Loop Reactors. *(48 ref.)*

Volume 14

Microbial Metabolism

1980. 39 figures. VII, 162 pages
ISBN 3-540-09621-3

Contents:
J.-C. Jallageas, A. Arnaud, P. Galzy: Bioconversions of Nitriles and Their Applications. *(143 ref.)*
Z. Řeháček: Ergot Alkoloids and Their Biosynthesis. *(192 ref.)*
R. V. Smith, P. J. Davis: Induction of Xenobiotic Monooxygenases. *(201 ref.)*
L. T. Fan, Y.-H. Lee, D. H. Beardmore: Major Chemical and Physical Features of Cellulosic Materials as Substrates for Enzymatic Hydrolysis. *(53 ref.)*
R. E. Spier: Recent Developments in the Large Scale Cultivation of Animal Cells in Monolayers. *(211 ref.)*

Volume 15

New Technological Concepts

1980. 39 figures, 27 tables. VII, 129 pages
ISBN 3-540-09686-8

Contents:
R. Seipenbusch, H. Blenke: The Loop Reactor for Cultivating Yeast on n-Paraffin Substrate. *(42 ref.)*
G. W. Pace, R. C. Righelato: Production of Extracellular Microbial Polysaccharides. *(140 ref.)*
T. Finocchiaro, N. F. Olson, T. Richardson: Use of Immobilized Lactase in Milk Systems. *(71 ref.)*
L. D. Bowers, P. W. Carr: Immobilized Enzymes in Analytical Chemistry. *(149 ref.)*

Springer-Verlag
Berlin Heidelberg NewYork

TELEPEN
Ø 157 518 Ø12